Quality Assurance for Scientists and Engineers

A practical guide

Online at: https://doi.org/10.1088/978-0-7503-5107-2

Quality Assurance for Scientists and Engineers

A practical guide

Steven Michael Judge
Brill, Buckinghamshire, UK

IOP Publishing, Bristol, UK

ISBN 978-0-7503-5107-2 (ebook)
ISBN 978-0-7503-5105-8 (print)
ISBN 978-0-7503-5108-9 (myPrint)
ISBN 978-0-7503-5106-5 (mobi)

DOI 10.1088/978-0-7503-5107-2

Version: 20231201

IOP ebooks

British Library Cataloguing-in-Publication Data: A catalogue record for this book is available from the British Library.

Published by IOP Publishing, wholly owned by The Institute of Physics, London

IOP Publishing, No.2 The Distillery, Glassfields, Avon Street, Bristol, BS2 0GR, UK

US Office: IOP Publishing, Inc., 190 North Independence Mall West, Suite 601, Philadelphia, PA 19106, USA

To my wife.

Contents

Preface

Many people study science but very few end up earning their living teaching and researching in a university. My education was excellent academically but did not prepare me for the world of work: it did not include many of the skills employers were looking for and how best I could contribute.

What is it like to be a scientist in a private company or a government institute? Perhaps the main difference with blue-sky research is that there is always a customer, someone who will say thank you for your work. The customers may be the people who are paying for your services or products, they may be colleagues in another part of the organization, or they may be the general public, the taxpayers who benefit from your work indirectly (even if sometimes they don't know it).

How you work day-to-day is also different. You bring your scientific knowledge and you ensure that this is always up to date, but your work may be more structured. Somewhere, a customer or someone acting as a proxy customer (a government department, for example) is paying; they want to know they are getting value for money and there tends to be more oversight, more checks. This takes some getting used to, but is not a bad thing—the last chapter explores how some aspects of this oversight may help address the so-called 'reproducibility crisis' in pure research.

Best practice for working in a structured way is described in 'quality assurance management systems', ways of working that draw on international experience over decades and that are captured in documentary standards such as ISO/IEC 17025 and ISO 9001. The term 'quality assurance' is often confused with 'quality control', the routine checks that you carry out on equipment. Quality assurance is far broader and impacts every aspect of work; a better term might be 'quality science' or 'customer-focussed science'. As a scientist, you will have to follow a quality system at some point in your career.

Hence the motivation for this book. The intention is to give a general introduction to the principles of quality assurance, to explain why they matter and how they help you and your organization. Pointers to where to go for further information are included.

The aim of this book is to capture a simple, pragmatic approach to quality systems from the point of view of a scientist. Case studies are included to illustrate key points. The book is based on practical experience of working in industry and in governmental organizations operating on the boundary between academia and industry. It is from experience of helping to set up quality systems, auditing other organizations and being audited, and from training courses, while working in a field where any errors could affect the health of many thousands of people.

Acknowledgements

The author wishes to thank everyone who contributed a case study to this book. Their support and encouragement was much appreciated. To the extent that the law allows, IOP Publishing disclaims any liability that any person may suffer as a result of accessing, using or forwarding the image(s). Any reuse rights should be checked and permission should be sought if necessary from Pixabay and/or the copyright owner (as appropriate) before using or forwarding the image(s).

Author biography

Steven Michael Judge

Steven Michael Judge is a technical writer living in the village of Brill in the United Kingdom. He studied physics at Corpus Christi College, Oxford University, before following a career in measurement science at the UK's National Physical Laboratory. He was awarded a PhD in nuclear physics from the University of Sussex in 1988.

He gained wide experience in metrology, working for a radio-pharmaceutical company, a nuclear industry site and the Laue-Langevin Institute in Grenoble, France. He learned about the benefits of following the structured approach to work that quality assurance management systems can bring, and also witnessed the challenges that scientists face adapting to this way of working. In the end, he concluded that the benefits far outweighed any problems, becoming a technical assessor for the United Kingdom Accreditation Service (UKAS) and working on an ISO committee developing new international standards.

Later in his career, he was appointed Director of the Ionizing Radiation Department at the International Weights and Measures Bureau (the BIPM) in Paris. He was charged with leading a team of dedicated scientists and technicians, working with colleagues from around the world to ensure the accurate measurement of ionising radiation for cancer therapy and environmental protection. In addition to his scientific work, he was responsible for the quality management system for the institute.

Now retired from full-time work, he continues to champion the benefits of quality assurance systems and works as a technical assessor for UKAS.

Chapter 1

Introduction—quality management systems

Quality management systems exist for one reason: other people care about what you do. Your colleagues, fellow scientists, or your customers are relying on your products or services, or the outcomes of your research. A quality management system is simply a structured way of working that has been proven over many years to help you meet your customers' needs. Quality management is often described as obvious common sense, the problem being that sense is sometimes not very obvious, nor is it particularly common.

A quality management system is a set of rules to follow in the workplace. The rules can be set out by local management, but most often are drawn from published documentary standards such as ISO 9001 or ISO 17025 (there will be more on these standards later). Following the theme of common sense, you might expect the rules to include:

(1) You must do what the customer or colleague is expecting
(2) You must be qualified and trained for your job
(3) You must use the right equipment and have the right facilities
(4) You must write down your results and get them checked
(5) You must identify the best way to carry out the work, write it down, and have it checked
(6) You must consider how accurate the results are

If you follow all these rules, there are perhaps three main benefits to you as a scientist. First, knowing that you are following the rules gives confidence to your customers or colleagues in your work. Second, writing down what you do helps you pass on your expertise to colleagues and to the next generation (a scientific paper rarely has enough detail to enable others to replicate the work). Third, the rules mean that you must be trained and must learn from people who have been there before, taking best practice from publications such as international standards. I have observed, with some frustration, nuclear physics experiments being set up with the students struggling with the same instrumentation issues others solved many years ago.

doi:10.1088/978-0-7503-5107-2ch1

1.1 A brief history of quality assurance

Many of the concepts of quality assurance management systems have their origins in the medieval guilds, the associations of craftsmen in Europe from the 11th century onwards [1]. The guilds were early examples of the advantages of division of labour, as specialisation helped to increase productivity and improve the standard of living. Apprentices joined a guild at an early age to learn their craft under the supervision of a master and were only promoted once they made an article of high quality, a masterpiece. Some guild masters acted as quality assessors, inspecting products from other guilds to ensure that they were of a high standard. Guilds were local monopolies and had much to lose if they lost their status through selling substandard products.

The guilds' monopoly status was eventually their undoing as it hindered trade, and their role disappeared during the industrial revolution (some guilds still exist, but as mutual support networks). The work of skilled craftsmen was replaced by factories, and the era of scientific management began. This philosophy was set out in the book *The Principles of Scientific Management* by Frederick Winslow Taylor in 1911 [2]. Taylor identified four principles: find the best method, train staff, monitor performance, and divide the work between people. During the 1930s, Walter Shewhart (of whom more later) built on this work applying statistical techniques to monitor, analyse, and improve work practices.

After the second world war, W. Edwards Deming is credited as the leading figure in the field [3]. Deming had a doctorate in mathematical physics from the University of Yale and became interested in applying Shewhart's techniques more widely. He was posted to Japan to help the country's recovery as industry faced the challenge to make best use of scarce raw materials. His work on quality and continuous improvement led to a new framework for management, including building quality in during manufacture rather than mass inspections, constantly improving, removing barriers to workmanship, and extensive training and education. To summarise Deming's method, success lay in quality rather than commercial short-term thinking.

Companies had their own quality systems during the 1970s. These systems were particularly important for the defence industry; quality had to be built-in as part of the manufacturing process rather than inspected-in by product testing. The United Kingdom decided to share best practice and devise a system that could be used by all industries, leading to the publication in 1979 of a national documentary standard, number BS5750. This was the first quality management system standard. Case study 1.1 describes an instrumentation company's experiences of adopting a quality assurance system.

The success of BS5750 and the increase in global trade led to the International Organization for Standardization forming a technical committee to write an international version of the standard. The first version of this standard, document number ISO 9000, was published in 1987. A family of quality assurance management standards has developed since, following the same principles but including sector-specific requirements such as ISO 17025 ('General requirements for the competence of testing and calibration laboratories') and ISO 17034 ('General requirements for the competence of reference material producers'). Some industries have retained their own approaches in parallel with the ISO series; for example,

Case study 1.1. The challenges and benefits of a quality assurance system for a small business.

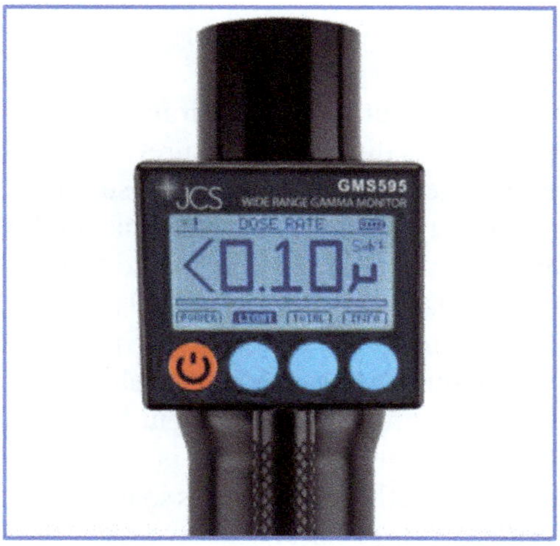

Our company, JCS Ltd, is based in the United Kingdom, and we design and build scientific instrumentation. We started our journey with quality assurance management systems with the early British standard BS5750. As a small company, we had to consider if the standard would be a help or a hinderance, but more and more customers were asking for accreditation to this standard, so it became essential that we adopted it.

At the time, we were struggling to complete the work we had in hand, and I had to overcome the perception of burdensome administrative steps that looked to be expensive and time consuming. I began to see that BS5750 was a very good way of organising and evaluating the working procedures of my company, as well as having many other advantages. When BS5750 was replaced by ISO 9001, there were clear improvements and we put the new standard into practice without hesitation.

To reduce the work needed in implementing the standard, we obtained copies of typical quality system procedures from a larger company and used them as templates (these days, templates can be found online). We shared the responsibility amongst ourselves, identifying the benefits to the company of the different sections. I feel this helped ensure that everyone in the company appreciated the reasons for the work.

Adopting ISO 9001 led to a new business model for the company during the 1990s. It occurred to me that our mature quality system could enable people to work together remotely—even before today's communications technology, there was no need to be co-located, and customers did not need to visit premises. Each member of staff knew what their role was and how it related to other members of staff, and new members of staff could easily learn their roles; one additional benefit for the team was a better work–life balance. Production of instruments was contracted out, and the quality system helped us to retain control, testing during production and checking instruments before shipping to the customer.

It required a new way of thinking, but adopting a quality system worked well for us: it reduced overhead costs, helped us meet the needs of our customers, and helped us become a successful instrumentation company in a competitive market.

John Caunt
Founder, JCS Ltd

quality assurance issues have been captured in regulations such as Good Manufacturing Practice in the pharmaceutical sector. The principles are, however, the same, drawing on the work of Deming and many others going back in history.

1.2 Quality infrastructure

Modern life is not possible without quality assurance. It is so important that there are national and international quality infrastructures—systems of public and private organizations, plus their policies and regulations—to support the quality of products and services. At work, you may not be aware that this quality infrastructure exists, as a good infrastructure becomes 'just the way we do things around here'.

A quality infrastructure has four parts (figure 1.1). First, all products, services, and scientific research rely on accurate, consistent measurements (the science of metrology). Second, written guidance is needed to help learn from the experience of others (documentary standards and best practice guides). Third, independent assessment is needed to show that your organization is following best practice (accreditation). Finally, in some cases quality assurance is so important that compliance with best practice has to be enforced in regulations to protect customers or the public (legal metrology). There are bodies at international and national levels responsible for these aspects of the quality infrastructure, and these bodies are described below.

1.3 Consistent measurement quantities and units

For all time, for all people

The 1799 motto of the metric system, Marquis de Condorcet

The bedrock of a quality infrastructure is the science of metrology—a consistent way to measure a physical, chemical, or biological quantity that is fit for purpose and accepted by colleagues or customers. This system is the metric system, the International System of Units (the SI [4]). It is perhaps one of the greatest achievements of mankind, as it has enabled so many advances in industry, healthcare, and environmental protection.

The story of metrology is long and often turbulent [5]. Historically, weights and measures were based on local traditions and there was resistance to change (the

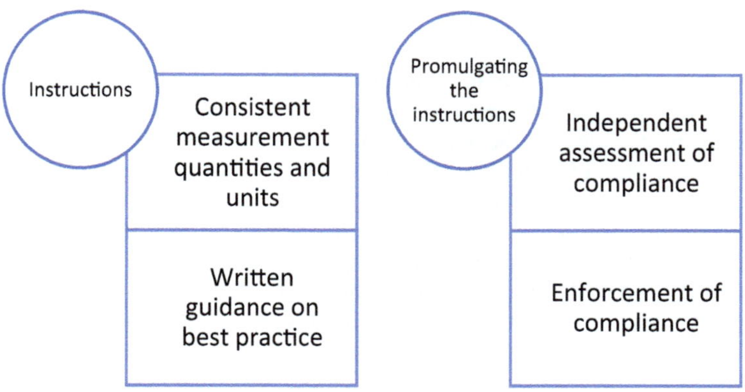

Figure 1.1. The four building blocks of a quality infrastructure.

resistance continues to this day in some areas). Many forward-thinking people recognized the need for a consistent system and contributed to the development of the SI. Some came from unlikely backgrounds: an English clergyman, the Reverend John Wilkins, is credited for publishing a concrete proposal for weights and measures based on scientific principles. In his essay 'Towards a real character and a philosophical language', published in 1668 by the Royal Society, he proposed a universal system of measurement based on a universal standard of length and a decimal system. It was not adopted at the time; the idea needed time and the fertile ground of the French Revolution to germinate.

In France at the end of the 18th century, there were some 800 different names for units with multiple definitions, resulting in an estimated 250 000 different units and in countless opportunities for cheating and exploitation. Fair weights and measures were one of the demands of the revolutionaries, a cry taken up by the Bishop of Autun who proposed a unified system in 1790, and the idea finally took root. A timeline of the evolution of the metric system is given in figure 1.2.

1668	The Reverend John Wilkins publishes a proposal for weights and measures based on scientific principles.
1790	Charles Maurice Talleyrand-Périgord, Bishop of Autun, proposed a unified system of weights and measures, based on the length of a pendulum with a period of 2 seconds. A Commission from the French Académie des Sciences accepted an alternative definition of the metre, based on the circumference of the Earth, as the length of the pendulum depended on the local value of acceleration due to gravity.
1795	The French Republic adopted the new units.
1799	Platinum artefacts representing the metre and the kilogram were deposited in the archives of the Republic and a law was passed implementing the new system.
1850s	World Fairs, including 'The Great Exhibition' held in London in 1851, showcased the products of the industrial revolution and highlighted the need for consistent measurements.
1869	The French government invited states to send representatives to form the International Metre Convention.
1875	The Metre Convention was signed, setting up the governing bodies for the metric system we see today.

Figure 1.2. The early steps towards a formal metric system. This timeline is drawn from records in the archives of the International Bureau of Weights and Measures, the UK Metric Association, and https://ukma.org.uk/what-is-metric/uk-progress/uk-metric-timeline/.

The advantages of the new metric system became so clear that it led to wide adoption, and an international treaty, the 1875 'Metre Convention', was signed. The Metre Convention set up the governance arrangements for the metric system and established a new international body, the International Bureau of Weights and Measures (the BIPM, as the French acronym is used by convention), as a permanent secretariat [6]. This body is one of the oldest international organizations in the world.

1.3.1 Governance of the international measurement system

The arrangements for the international measurement system have broadly been unchanged since the Metre Convention was signed[1]. The governing body is the General Conference on Weights and Measures (the CGPM); this is a meeting of scientists and diplomats from all the member states[2] that have acceded to the Metre Convention. States can also apply to become an associate state with the intention of acceding to the Metre Convention at a later date; representatives from associate states can attend meetings of the CGPM but are not allowed to vote. In 2023, the members and associates represent 100 states and economies. The CGPM is the body that takes all the policy decisions about the SI.

One illustration of the impact of the work of the CGPM is the measurement of mass, an issue that was covered widely in the press[3]. All measurements, from the weight printed on the packaging of foodstuffs in your local supermarket to high-precision manufacturing, used to be based on a single artefact, a platinum–iridium cylinder 39 mm in diameter and 39 mm high. This artefact is called the 'International Prototype of the Kilogram' (IPK), and it had a mass of exactly one kilogram by definition. The IPK and a set of identical copies were manufactured by Johnson, Matthey & Co. in 1878–79; the IPK was placed at the International Bureau of Weights and Measures in Paris, and the official copies were distributed to national laboratories worldwide. This dependence on artefacts was risky; the surface of the artefacts can become contaminated, so they must be cleaned, and the mass could change. Gradually over the next century, small discrepancies were observed between the IPK and the copies—there was a weakness in the system for the measurement of mass. Following three decades of research [7], alternative approaches to realise a standard kilogram were devised. The CGPM voted in 2018 (figure 1.3) that the kilogram would be redefined and would be based on the Planck constant rather than an artefact; the benefits from this decision will be felt over the next few decades.

The CGPM only meets once every four years, so a support structure is needed. The work of implementing the CGPM's policies is the responsibility of the

[1] Further information is available on the website of the International Bureau of Weights and Measures: https://www.bipm.org/en/.

[2] More formally, member states are known as 'States Parties to the Metre Convention'.

[3] The evolution of the SI units is described on the website of the International Bureau of Weights and Measures: www.bipm.org/en/history-si.

Figure 1.3. The meeting of the CGPM in 2018 and the historic decision to replace the last artefact—the International Prototype of the Kilogram. Photograph: author's own.

International Committee for Weights and Measures (the CIPM). The CIPM promotes uniformity of measurements, either directly or by submitting draft resolutions to the CGPM. The CIPM has eighteen members of different nationalities appointed by the CGPM, who are leading metrologists and senior managers from national metrology institutes (of which more below). The CIPM is advised on the issues faced by different fields of measurement by specialist committees (Consultative Committees) and their Working Groups.

The Consultative Committees are where the bulk of the scientific work is carried out. These committees are a forum for representatives from metrology institutes and international organizations to exchange information and advance the science of metrology. Each Consultative Committee has its own strategic plan, aiming to encourage the development of new methods and seeking to improve global comparability of measurements by organizing comparisons of measurement standards, creating opportunities for collaboration, and using a common vocabulary to improve communication. Consultative Committees can set up less formal Working Groups and Task Groups, bringing together experts to investigate specific topics. All of the Consultative Committees have the same overarching aim—to lead a globally recognised system of standards, methods, and facilities for measurements in their field.

The minutes of Consultative Committee meetings are published on the website of the International Bureau of Weights and Measures and can be followed back to the beginnings of the field—the committees bring consistency globally and over a long time period. Each Consultative Committee is chaired by a member of the CIPM who reports back to the CIPM regularly. Membership in a Consultative Committee is restricted to national metrology institutions recognised as the most expert in the field; this does not mean that the institution must be old (although many are, dating back more than a hundred years), but it must have a good record of research, scientific publications, and successful participation in comparisons of measurement standards. International organizations that share the same mission may also be invited to participate in meetings.

A national metrology institute seeking the status of membership in a Consultative Committee has to apply to the CIPM to send an observer to the meetings to engage more closely with the international system. When the institute thinks it may meet the criteria for full membership, it can apply again to the CIPM; the institute may be asked to give a presentation to the Consultative Committee on its work. The CIPM will then make the decision on membership after consulting the Consultative Committee chair.

The CGPM, the CIPM, and the Consultative Committees are supported by a permanent scientific secretariat, the International Bureau of Weights and Measures (BIPM). The BIPM is based in Sèvres on the outskirts of Paris, where it has been located since its foundation in 1875. The original purpose of the BIPM was to house the artefacts that set the international standards, the IPK and the prototype metre. It was found that it made scientific and economic sense for the BIPM to be the focal point for other fields of measurement, and the BIPM laboratories now provide services in mass, electrical, chemical, and ionizing radiation metrology. The BIPM is also responsible for the Coordinated Universal Time (UTC), the time standard by which the world regulates clocks and time. Another key role of the BIPM is to provide knowledge transfer and capacity building activities, including e-learning, webinars, and workshops.

The structure and rules for overseeing metrology may seem formal, but the rules are intended to ensure all existing and future members are treated equitably. The BIPM publishes the procedures on its website.

1.3.2 National metrology institutes and designated institutes

The arrangements for the international measurement system have been in place since 1875, but you are more likely to encounter the system at the national level. At the end of the 19th century, states realized the importance of measurement science to the economy—in the United Kingdom, this resulted in the foundation of the National Physical Laboratory in Teddington 'to bring scientific knowledge to bear practically upon our everyday industrial and commercial life' in the words of the Prince of Wales during the opening ceremony in 1902. Establishing the National Physical Laboratory was the first time that the UK government had taken part in scientific research. Other countries had similar institutes (the Physikalisch-Technische Reichsanstalt, the forerunner of the Physikalisch-Technische Bundesanstalt, was founded in Germany in 1887), and by 2022 there were 97 such national metrology institutes (NMIs) worldwide.

Each state is represented by a single metrology institute that takes the lead role. However, the state can delegate responsibility for specific fields of metrology to other laboratories—these laboratories are called designated institutes (DIs). This is common in the field of ionizing radiation metrology, which needs specialist equipment and safety precautions. DIs hold national measurement standards and provide services to customers; they can be commercial companies but must not use their status as a DI to gain advantages over their competitors. There is a cost involved, as DIs have the same obligation to participate in the international system as NMIs.

1.3.3 Regional metrology organizations

There is one more layer in the international measurement system (see figure 1.4). National laboratories work together in local groups called regional metrology organizations (RMOs). There are currently six RMOs: AFRIMETS (Africa), APMP (Asia-Pacific), COOMET (Euro-Asia), EURAMET (Europe), GULFMET (Gulf states), and SIM (North and South America).

Similar to the international system, RMOs have their own general assembly to decide policies, supported by a permanent secretariat and technical committees or working groups covering the different technical fields. Membership is inclusive and extends to all NMIs and DIs in the region. RMOs aim to support all their members to meet the needs of their states, and they have active knowledge transfer and capacity building programmes.

The role of RMOs is wide-ranging and includes engaging with industry and other stakeholders to ensure that the science of metrology is evolving to meet changing needs. Some RMOs have their own funding mechanisms and offer opportunities for NMIs, DIs, industry, and academia to collaborate on research projects.

RMOs have a key role to play in ensuring the quality of the work of the national laboratories by appointing scientists to carry out independent peer reviews of the capabilities of laboratories both within their region and in other regions. RMOs also organize and fund research projects in the field of metrology.

To summarize, the international measurement system is overseen by representatives from states who have signed an international treaty called the Metre Convention, supported by a permanent scientific secretariat (the BIPM), and promulgated at the national level by NMIs and DIs supported by RMOs. The structure supports existing metrology institutes but also provides a route for a new national laboratory to gain knowledge and experience, progressing from

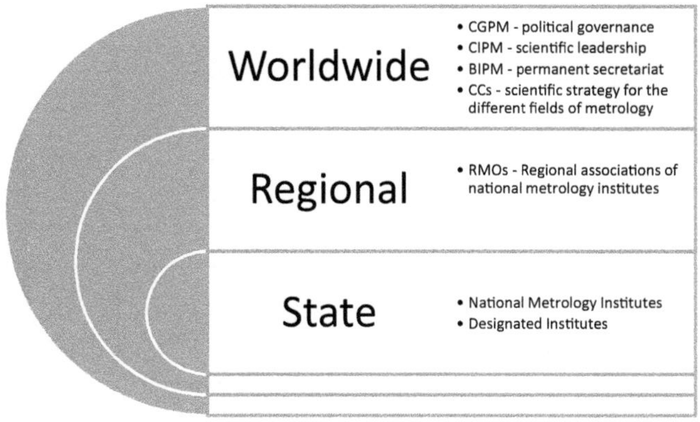

Figure 1.4. Summary of the structure of the international measurement system.

membership in the regional organization to full membership in the international Consultative Committee.

The next aspect to consider is how this system affects your work from a scientific perspective, whether you are working in industry or in academia. All of the basic and general concepts of measurement science are set out in a publication referred to as the VIM (the French acronym for *International Vocabulary of Metrology*) [8]. The three most important concepts are traceability, equivalence, and measurement uncertainty.

1.3.4 Traceability and calibration

NMIs are the custodians of a set of primary measurement standards that are the reference points for measurements for the state. In the past, the primary reference standards would have been artefacts—exact copies of the IPK held by the BIPM, for example. These days, primary measurement standards are created ('realized' to use metrologist's jargon) based on physical principles. The definitions and principles of the seven base units are summarized below in figure 1.5, taken from the SI brochure [4]. Twenty-two other named units are derived from these seven base units.

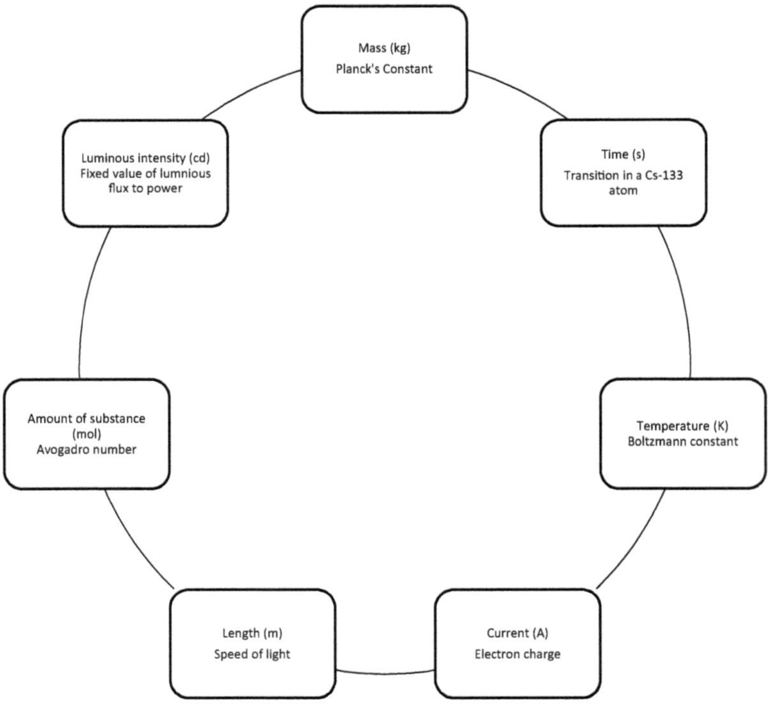

Figure 1.5. The seven base units of the SI. Since 2019, all of the units are based on fundamental constants, a property of nature, or a fixed value.

In daily life, measurement is taken for granted. Whether it is the weight of the goods you purchase at the supermarket, the volume of petrol dispensed into the tank of your car, or the measurement of airborne pollution, the value stated is simply the ratio between what you are trying to measure and a standard held by an NMI. This link between your measurement and the national measurement standard is made by calibrating the instrument you are using. In the language of the VIM, calibration is the 'operation that, under specified conditions, in a first step, establishes a relation between the quantity values with measurement uncertainties provided by measurement standards and corresponding indications with associated measurement uncertainties and, in a second step, uses this information to establish a relation for obtaining a measurement result from an indication' [8, p 28].

The proof that an instrument has been calibrated is a written (or electronic) record, a calibration certificate; sometimes you will see a label attached to the instrument that confirms when it when it was last calibrated. In most cases, instruments are not calibrated by the NMI itself but by a company with specialist instrumentation or artefacts that have themselves been calibrated. There are exceptions in particularly sensitive fields of measurement—for example, the instruments used to check the radiation doses for cancer therapy are calibrated by the NMI—but normally there are several calibration steps between you, the user, and the national measurement standard.

This documented chain of calibration is called 'traceability'. The VIM defines traceability to be the 'property of a measurement result whereby the result can be related to a reference through a documented unbroken chain of calibrations, each contributing to the measurement uncertainty' [8, p 29]. The key point is that each calibration must be documented—an instrument that you use for a measurement may give an accurate reading, but you must be able to prove it.

1.3.5 Equivalence and the CIPM Mutual Recognition Arrangement

Traceability means that any measurements you carry out can be linked back to a particular national standard. The next challenge is to demonstrate that the national standards are identical, so that you can have confidence that your measurements match those carried out by your peers worldwide, and you can trade with other countries or know that healthcare or environmental protection measures are the same.

This important concept is called 'equivalence', and how this is shown in practice is captured in a document called the CIPM Mutual Recognition Arrangement (CIPM MRA) [9]. The CIPM MRA was published in 1999 and describes how equivalence is demonstrated in an objective way. The CIPM MRA has three aims: to establish the degree of equivalence of national standards in a quantitative way, to enable the mutual recognition of calibration and measurement certificates between countries, and to set a secure technical foundation for trade and regulations. One of the drivers for the CIPM MRA was to eliminate technical barriers to trade.

The first step to demonstrate equivalence is that NMIs must participate in international comparison exercises of their standards. The worldwide exercises are overseen by the relevant Consultative Committee and are led by an NMI (called the pilot laboratory). The exercises follow strict rules set out on the BIPM website: participation is limited to laboratories with the highest level of expertise in the realization of standards, NMIs are not allowed to change their values once submitted, results are kept confidential from other participants, and the outcome from the comparison exercise must be published. The report is peer-reviewed by the Consultative Committee before it is accepted for publication. The approach is thorough, but it risks excluding smaller or newer NMIs, so RMOs arrange similar exercises open to all the NMIs in their region. Whether the comparison is organized nationally or regionally, participation is not taken lightly, as the reputation of the NMI can be at stake.

The mechanics of the comparison exercise depends on the field of measurement. For example, in radioactivity, national standards (in the form of activity in solution (Bq g^{-1})) may be compared using a high-precision device held at the BIPM or by a pilot laboratory dispensing a set of identical ampoules that are distributed to participating laboratories for measurement. Other fields of measurement have other approaches, such as circulating an instrument that is calibrated in turn by each of the participating laboratories. The project plan for the comparison, including details of what is expected of each laboratory, is published in the form of a protocol on the BIPM website.

Whatever technical approach is taken, a reference value called the Key Comparison Reference Value (KCRV) is calculated for the quantity of interest. The KCRV is considered to be a close approximation to the 'true' SI value—in other words, a consensus value of the international measurement standard. The CIPM MRA does not define how the KCRV is to be calculated; this is left to the Consultative Committee to decide. The obvious approach is to take a weighted mean of the results, but the calculation is more difficult than it may seem at first—how one can decide that a result is an outlier and should be omitted from the calculation is difficult, particularly when that decision impacts a national standard. Many other issues arise in practice; for example, there can be subsets of the data that are consistent but discrepant from each other, and much has been published in the scientific literature on the advantages and disadvantages of different methods (for example, see [10, 11] and references therein). In the end, it is the role of the Consultative Committee to decide for its field of measurement, and the approach is agreed and published in the protocol in advance.

Having calculated the KCRV for the comparison exercise, the 'degree of equivalence' for each participating laboratory is expressed as two terms: the deviation from the KCRV and the uncertainty in the deviation at the 95% level of confidence. The results from a comparison exercise must be confirmed by the Consultative Committee and published in a database called the Key Comparison Database (KCDB). An example from an international comparison of national standards of a radioactive isotope of cobalt (^{60}Co) is shown in case study 1.2. If an NMI finds that its result is discrepant, the NMI can review its procedure, correct the error, and arrange a bilateral comparison exercise with one of the original

Case study 1.2. The measurement of ^{60}Co.
Nuclear power may have an important role in ensuring a secure energy supply. All nuclear facilities discharge small quantities of radioactivity into the environment, and the public needs confirmation that the quantities are within safe limits. The international measurement system ensures that measurements of radioactivity in the environment can be relied on.

Primary standards of ^{60}Co

- National Metrology Institutes work independently to realise primary standards of ^{60}Co using specialist instrumentation.

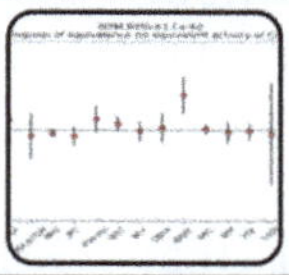

Equivalence of primary standards

- The national primary standards are compared, using a service offered by the BIPM.

Traceable reference materials

- Commercial suppliers manufacture reference materials, calibrating their instruments in a way that is traceable to a primary standard.

End users

- Laboratories in the nuclear industry measure the ^{60}Co content of environmental samples, using instruments calibrated using the reference materials. Results are reported to regulatory bodies and are published.

The photograph of a coincidence counter (top image) is reproduced courtesy of the National Physical Laboratory, UK. The partial results shown from a comparison exercise are extracted from the BIPM Key Comparison Database. The photograph of reference materials is reproduced courtesy of Eckert & Ziegler Analytics. The final photograph is by Kilian Karger on Unsplash.

participants to show that the (new) national standard is equivalent, but both results remain in the public domain and are published on the BIPM website. Further details on the different types of international comparison exercises and the procedures followed are given in guidance documents published by the BIPM.

Each NMI is responsible for deciding internally the impact of the result from a comparison exercise on its national standard. In some cases, the NMI will use the result to justify the assertion that its standard is 'correct'; some NMIs may apply a correction factor and adopt the international value (the KCRV) as the national standard.

Key comparisons can be complex and time consuming to organize, and they are intended to be for the main quantities. Consultative Committees or RMOs can also organize 'supplementary comparisons' to address specific needs of the sector; for example, the measurement of ^{60}Co in foodstuffs might be covered by a supplementary comparison to complement the comparison of primary ^{60}Co standards. Another format for comparisons is a bilateral comparison between two NMIs; this approach can be used if an NMI wants to join a comparison exercise that has finished by linking to a participating laboratory. Bilateral comparisons are also used if the result submitted by the laboratory was discrepant, changes have been made to procedures, and the laboratory wishes to demonstrate that its standards are now equivalent.

It is rarely possible for key or supplementary comparisons to cover all the different quantities and ranges in a scientific field in a reasonable timeframe. Consultative Committees have to choose priorities, and some have developed guidelines that accept that successful participation in a comparison exercise for one quantity is accepted as proof of competence for a restricted set of other quantities.

To summarise, an NMI has to demonstrate equivalence of its national standards through publication of the result from a comparison exercise. This provides scientific evidence, but it could be difficult for a customer looking to establish traceability to a national standard to find an NMI that offers the service and then to confirm global equivalence. The customer-focussed aspect of the CIPM MRA is that NMIs are expected to publish their services (referred to as Calibration and Measurement Capabilities (CMCs)) in the KCDB database hosted by the BIPM; it is easy to search the database to find which NMIs offer the service wanted.

Publication of an NMI's capabilities in the database is not automatic. The NMI must submit a claim for a CMC using a software package on the KCDB. The claim must include details of the service such as the quantity, range, and measurement uncertainties along with evidence that the NMI has the competence to carry out the service. The main publication accepted as evidence is the result from successful recent participation in a comparison exercise, but other evidence is also considered, including the publication of scientific papers, contributions to research projects, results from participation in a closely related comparison exercise, and the outcome of on-site peer assessments. Consultative Committees are responsible for defining the evidence that can be taken into account in their field.

The claim and the supporting evidence are reviewed by experts in the field from the RMO; if accepted, the claim is submitted for a second review by experts from another RMO. Only after passing these two review steps can the CMC be published in the KCDB.

These two steps—successful participation in a comparison exercise and independent review of competence by experts—give compelling evidence of the equivalence of national standards. There is one further requirement of the CIPM MRA, a requirement to ensure that NMIs maintain their competence: the NMI must follow the ISO 17025 quality standard and be subject to independent audit or peer review. This ensures that NMIs must have trained, experienced staff, have the right equipment and facilities, and follow scientifically rigorous procedures (so, for example, how an NMI uses the results from comparison exercises is subject to independent scrutiny).

In summary, there is an internationally agreed mechanism (the CIPM MRA) to demonstrate that national standards are equivalent. The mechanism requires NMIs to take part in comparison exercises, to have a quality system, and to have their services subject to independent review. Nothing is left to chance; there are independent checks and balances throughout. This may seem a complex infrastructure and a heavy burden, until one considers the importance of consistent measurements worldwide and the impact on trade, health, and the environment.

1.3.6 Measurement uncertainty

The third essential concept in metrology is the uncertainty of measurement. This topic is rarely covered in degree courses in detail, but it impacts all scientific studies. To explain why uncertainty is so important, here are first some definitions in the formal language of metrology [8]:

Quantity: the 'property of a phenomenon, body, or substance, where the property has a magnitude that can be expressed as a number and a reference' [8, p 2]. For example, length, mass, and time are all 'quantities'. A 'measurand' is the quantity intended to be measured.

Measurement unit: a 'real scalar quantity, defined and adopted by convention, with which any other quantity of the same kind can be compared to express the ratio of the two quantities as a number' [8, p 6]. Measurement units (for example, metre, kilogram, time) have symbols assigned by convention. Measurement units can be 'base units', units that cannot be expressed in terms of other units, or 'derived units' which can be expressed in some combination of base units.

Quantity value: the 'number and reference together expressing magnitude of a quantity' [8, p 12]—in other words, a measurement in the form <number> <unit>. The 'true quantity value' is the true value of whatever it is one is measuring, the exact length of an artefact, for example.

In any measurement, one is seeking to determine the true quantity value, but in practice this is unique and unknowable. Measurements are always perturbed—changes in the laboratory temperature changing the readings, problems with reading a scale on a ruler, slight variations in the procedure one is following, etc—all mean that the result from a measurement has some uncertainty associated with it, and one can only say that 'the true value lies somewhere between x and y with a certain level of confidence'. Again, in the language of the VIM:

Measurement result: the 'set of quantity values being attributed to a measurand together with any other available relevant information' [8, p 19].

This means that a measurement result is normally expressed as a single measured quantity value and a measurement uncertainty. Of course, in daily life and in many fields, the measurement uncertainty is rarely seen because it is considered to be negligible. The uncertainty in the measurement is still there; it is just that someone else further up the chain of traceable measurements has worried about this on your behalf so you do not have to. As a scientist, this may be your job.

Measurement uncertainty: 'non-negative parameter characterizing the dispersion of the quantity values being attributed to a measurand, based on the information used' [8, p 25].

In other words, a measurement result for a quantity has three aspects:

<the value—a number> <the unit—a symbol> <a way to describe how close the value is to the true value—the uncertainty>

The measurement uncertainty is usually expressed as the standard uncertainty at a given level of confidence, so that the user knows the likely range in which the true value lies. How uncertainties are estimated is described in more detail in chapter 2. Measurement uncertainty should not be confused with an 'error', which is a mistake, something that has gone wrong (although you may see the term 'error' used in some old publications).

A lot of importance is attached to measurement uncertainty in metrology. Estimating the uncertainty enables the results from different laboratories or groups to be compared fairly, helps customers to interpret the data, and enables measurements to be improved by identifying what is causing the uncertainty. It also enables the measurement uncertainty to be omitted further down the traceability chain if there is evidence that it is negligible for the application. Omitting the uncertainty sometimes leads to problems: for example, if a nuclear power station changes the technique used to measure discharges of radioactivity into the environment, a sudden step change in the results sent to the regulator can indicate something has gone wrong, whereas in reality the true value is unchanged and the old and new results are in agreement, taking into account the measurement uncertainty. Another example of the value of estimating measurement uncertainty is shown in case study 1.3.

Case study 1.3. How uncertainty calculations improved testing turbofan jet engines.
A major problem in the aerospace industry is the determination of nozzle thrust from a turbofan jet engine. Measurements of thrust are needed to assess the performance of new designs and to test engines returning to service after repair or overhaul.

Test facilities for indoor and (right) outdoor data acquisition

A major problem in the aerospace industry is the determination of nozzle thrust from a turbofan jet engine. Measurements of thrust are needed to assess the performance of new designs and to test engines returning to service after repair or overhaul.

Nozzle thrust can be derived from measurements of exhaust flow but requires multiple measurements of pressure and temperature. Sensors must be positioned in the flow to take the readings, but the sensors perturb the flow. A new technique, sonic anemometry, using acoustic sensors, has been developed to avoid this problem.

Transmitters and receivers are positioned on opposite sides of the exhaust flow; the time for an acoustic signal to pass from the transmitter to the receiver depends on the distance, the speed of sound, and the flow velocity. The flow velocity can therefore be calculated indirectly from acoustical measurements, with no sensors inside the exhaust.

The key to adopting sonic anemometry in practice was detailed calculations of the measurement uncertainty. The approach described in the *Guide to the Expression of Uncertainty in Measurement* (GUM) [12] was adopted; the uncertainties associated with the geometry, the properties of the gas, and measurement of the time were propagated to determine the measurement uncertainty of the exhaust flow. The calculations showed that it was possible to reduce the uncertainty further by careful positioning of the transmitter and receivers.

This is just one example of how the methods described in the GUM enable innovation—there are many other examples of the impact in healthcare, environmental protection, and industry.

(This work constituted part of project 17NRM05 EMUE, which received funding from the EMPIR programme co-financed by the Participating States and from the European Union's Horizon 2020 research and innovation programme.)

Professor Maurice Cox, Emeritus Fellow, National Physical Laboratory, United Kingdom

In the same way that the concepts of traceability and equivalence harmonized measurements worldwide, the estimation of uncertainties has also been harmonized. Work on this challenge started in 1977, when the BIPM consulted NMIs and formed a Working Group to arrive at a consensus. There was wide interest in the work, and the International Organization for Standardization along with other international organizations (CIPM, the International Electrotechnical Commission, the International Organization of Legal Metrology, the International Union of Pure and Applied Chemistry, the International Union of Pure and Applied Physics, and the International Federation of Clinical Chemistry) developed the definitive guidance (*The Guide to the Expression of Uncertainty in Measurement*, or GUM [12]) which was published in 2008. The GUM has brought international consensus to the evaluation and expression of uncertainty in measurement.

The GUM had three objectives. First, the guidance had to be applicable to all kinds of measurement. Second, it had to be possible to use the uncertainty in measurements on which the first measurement depends (for example, the user must be able to take the uncertainty stated on a calibration certificate for an instrument and combine it with other components of uncertainty in their measurement). Third, it had to be possible for the user to calculate a range, at a stated confidence interval, in which the 'true value' lies, so the user could state the uncertainty depending on the risk for their application. These thoughts led the authors of the GUM to adopt an approach that treats all components of uncertainty in the same way.

The GUM requires all components to be estimated as a 'standard uncertainty'—the familiar 'standard deviation' from statistics. The guidance then classifies uncertainties as 'Type A' or 'Type B', where 'Type A' uncertainties are estimated by replicating measurements and 'Type B' uncertainties are estimated by other methods (for example, taken from the calibration certificate for the instrument used). The components of uncertainty are then combined to obtain a realistic value for the uncertainty. Previous methods for estimating uncertainties sometimes deliberately overestimated the total uncertainty: as well as making it difficult to compare measurements in different laboratories, this approach could lead to purchasing expensive high-accuracy equipment that was not needed or the unnecessary rejection of products or services. There were sometimes good reasons for overestimating uncertainties, for example, in the characterization of pharmaceuticals when you might want to take a cautious approach, but the authors of the GUM considered that the benefits outweighed these concerns.

Old habits die hard, and you may still come across the terms 'random' and 'systematic' uncertainties rather than Type A and Type B uncertainties, and you may encounter different approaches to estimating and combining uncertainties. However, the approach described in the GUM is mathematically rigorous and is the definitive method to adopt.

1.3.7 Harmonization

The international measurement system (the SI) and the concepts of traceability, equivalence, and uncertainties calculated following the GUM work together to provide a harmonized global system. Harmonized does not mean that everyone agrees with it all the time—popular resistance to the metric system continues to the present day in some countries, and methods for calculating uncertainties are debated in scientific committees—and you may not agree with some approaches yourself.

However, the benefits of using the SI and working in the same way as other scientists worldwide must surely outweigh any disadvantages. It is the best hope we have of realizing Condorcet's vision of a measurement system 'for all people, for all time'.

1.4 Written guidance that describes best practice

A quality infrastructure also needs guidance on best practice, a way to capture the expertise of colleagues who have faced the same challenges and found solutions that work. There are several international organizations that exist to develop written guidance. The largest of these standards bodies is the International Organization for Standardization (ISO); its origins go back to a group of engineers who founded the International Standardization Association in 1947 to write documents to share their knowledge. The documents were called 'recommendations'; they have been called 'international standards' since 1971 (not to be confused with reference standards, physical artefacts or instruments used for calibration). It is an independent, nongovernmental organization, funded by subscriptions paid by national standards bodies (see below). It also relies on the income from sales of standards, so understandably ISO defends its copyright with some vigour.

The documents were focussed on technical topics, but this was extended to more generic standards, such as the well-known ISO 9000 series on quality management systems that help organizations meet customer requirements. This series was first published by ISO in 1987, and there are now more than one million organizations accredited to this standard (you will see the ISO 9000 symbol on the websites or marketing information of many companies you deal with every day).

Nowadays, ISO aims to facilitate trade in goods and services, establishing provisions for common use. A typical standard will set out the scope[4]—the topic it covers and what it doesn't cover, so that the reader is clear on what to expect. The terms used in the standard are explained in some detail to establish a common vocabulary that matches other related standards. Guidance is then set out; the level of detail depends on the standard, sometimes going into step-by-step instructions and sometimes summarizing the guidance in a few carefully written sentences. The language used in the sentences is important: 'shall/shall not' means the instructions are compulsory, 'should/should not' means they are recommended, 'may/need to' means the standard gives permission to do something, and 'can/cannot' describes a possibility.

The text can be 'normative', meaning that you must follow the instructions to comply with the standard, or 'informative', which is not compulsory—other approaches are allowed. Additional information is given in annexes. The amount of text rarely matches the amount of effort needed to implement the standard; one sentence can imply a lot of work once its contents are unpacked.

From a commercial point of view, following a common approach saves money as it avoids reinventing what others have already done, ensures parts are interchangeable, and enables products or services to be sold in different countries. Drawing on global expertise can also help reduce waste, improve processes, and help an organization launch new products or enter new markets. In this digital era of artificial intelligence, robotics, automation, and the 'internet of things', standards can help generate trust in novel approaches, encourage the widespread adoption of technology, and ensure interoperability. The value in bringing experts together to write the standards should not be underestimated.

From a macroeconomic point of view, documentary standards ensure that markets are not dominated by one company or by one country, contribute to economic efficiency and productivity, and enable stakeholders to participate in setting standards improving consumer choice. They can help drive the introduction

[4] Further information on documentary standards can be found on the ISO website: https://www.iso.org/.

of new technology—publication of an ISO standard indicates that there is general expert consensus, reducing the risk of adopting a new approach.

There are currently more than 20 000 published ISO standards covering almost every aspect of our daily lives, from road safety to medical devices. ISO has developed a methodology to assess the impact of standards for an organization, and there have been many studies of the benefits, supported by case studies[5]. The British Standards Institute estimates that standards contribute more than £8 billion to the UK economy[6]. Perhaps the most famous standard was first published in 1968 concerning the design of freight containers for shipping—the standard was so well used that the shipping containers are now called 'ISO containers'. Suffice it to say that you will encounter standards throughout your scientific career, as tools to help your work, as references when you are purchasing equipment, as requirements from your customers, colleagues, or regulators, or as a consequence of your organization's policy to improve efficiency.

The way that ISO (and other standards bodies) operates is interesting, as it means that you may have an opportunity to contribute to a standard by commenting on its contents or by helping to draft the document. ISO has few resources of its own; it has a small central secretariat of about 150 staff based in Geneva. The work is overseen by a General Assembly attended by representatives from the member bodies and the smaller ISO Council of twenty member bodies and the principal officers of ISO. A Technical Management Board reports to the Council. The bulk of the work is carried out by the technical committees, subcommittees, and working groups composed of volunteers funded by their own organizations, by national standards bodies, or, sometimes, self-funded. These committees and working groups are responsible for writing and maintaining the standards.

Once a working group has accepted the draft standard, it is circulated to national standards bodies for voting and comment. Any comments on the text have to be justified, and the national standards body must suggest alternative text (comments such as 'I don't agree with this' are ignored by the working group). There are several voting and commenting cycles before the standard is published; the whole process takes about three years, sometimes much longer.

Although this section has referred mainly to ISO as it is the largest publisher of voluntary documentary standards, other standards bodies operate in a similar way with a different scope of work; the largest of these are the International Electrotechnical Commission (IEC) and the International Telecommunication Union.

To summarize, documentary standards are developed by experts in the field, peer-reviewed by colleagues on the working group, and peer-reviewed again by national standards laboratories. A standard captures the collective knowledge and expertise of scientists and engineers worldwide.

---

[5] See *The Economic Benefits of Standards*, available on the ISO website: https://www.iso.org/publication/PUB100288.html.

[6] Available on the British Standards Institute website: https://www.bsigroup.com/en-GB/standards/benefits-of-using-standards/research-reports/.

1.4.1 National standards bodies

None of the international standards could be written without the support of national standards bodies which are appointed by governments to represent the country's interests. The oldest national standards body is the United Kingdom's British Standards Institute (BSI) which was founded in 1901. The national bodies mirror the work of the international standards institutes, with committees that match the international committees. The BSI, for example, currently has more than 1000 committees in which some 12 000 people contribute.

One of the roles of the national bodies is to engage with all the stakeholders who may have an interest in the standards, including industry, government bodies, trade associations, academia, and consumers. There are two main ways for you to get involved. First, you can contribute to the development of standards using your local body's standards development portal—you can comment on drafts and even suggest new standards that are needed (comments on standards are welcomed; all too often valuable feedback is received only after the standard has been published and it is difficult to make rapid changes).

Second, if you have the relevant expertise, you can apply to the standards body to become a committee member; you may also be nominated by your organization, professional body, or university. Membership of the local committee can lead to you being nominated to represent the standards body at international meetings.

To summarize, national and international documentary standards are best practice guides that capture years of experience. Standards are a very useful resource you can call upon for your work, and they can help you avoid repeating errors. There are tried and tested procedures at both national and international levels to write, review, and publish the standards to ensure that there is general consensus. Getting involved in writing the standards is personally very rewarding; although you don't get your name attached to the publication as you do for a scientific paper, you have the satisfaction of knowing that your work will be used worldwide and will make a real difference. You also gain opportunities for professional development and build a network of contacts in your field.

1.4.2 Good practice guides

There is a third level of advice you can draw on. Professional bodies, trade associations, industry bodies, and national and international organizations often publish good practice guides, so it is always worth searching on the web for the topic you are interested in (for example, the United Kingdom's National Physical Laboratory publishes more than 100 good practice guides on its website[7]). These guides are written by experts in the field and can be very helpful, but are not subject to the same formal scrutiny applied to ISO or similar standards. However, they can be published faster than a formal standard and are usually free of charge.

[7] Measurement Good Practice Guides are available on the National Physical Laboratory website: https://www.npl.co.uk/.

1.5 Accreditation

1.5.1 International level

Complying with a documentary standard can be expensive and time consuming; the benefits of advertising compliance may be clear, but the temptation for an organization to cut corners to save cost may be strong. In the real world, an independent way is needed to check that an organization is following all requirements of the standard. The responsibility for this task falls to the network of national accreditation bodies. The national bodies accredit organizations within their countries, largely by conducting audits to check that the standards are followed. This has been a successful approach, and by 2021 there were 85 000 laboratories accredited worldwide, with the national accreditation bodies providing a stamp of approval of these laboratories.

The national accreditation bodies themselves need an international stamp of approval, to ensure that laboratories in different countries are being accredited to the same criteria, ensuring that the audit criteria are applied with equal rigour. Only when this happens will one country accept test and calibration work carried out in another country, facilitating trade, reducing the need to duplicate work across borders, reducing the risk of accepting lower quality goods and services, and supporting regulations.

There are several international organizations whose role is to provide this stamp of approval of national accreditation bodies, and to coordinate and support their work. One such organization is the International Laboratory Accreditation Cooperation (ILAC)[8] which was established in 1977 to cover topics such as calibration, testing, inspection, reference materials, and proficiency testing. National accreditation bodies are accepted as members of ILAC following stringent peer evaluation, with a team of auditors visiting the national body and witnessing assessments in progress. Members commit to complying with an ISO/IEC standard on conformity assessment (ISO/IEC 17011) and agree to assessing laboratories to ISO/IEC 17025 (test and calibration), ISO/IEC 17043 (proficiency test providers), ISO/IEC 17034 (reference material producers), or ISO/IEC 17020 (conformity assessments). Having passed the peer evaluation, members sign the ILAC MRA, reducing technical barriers to trade between countries. By 2022, 106 accreditation bodies had been accepted as signatories.

In practice, ILAC provides a mechanism for exchanging information on accreditation and developing good practice. It provides assistance to accreditation bodies seeking to become members, and it promotes the concept of accreditation to industry, governments, regulators, and consumers.

ILAC covers laboratory work; there is a sister organization called the International Accreditation Forum (IAF) which operates in a similar way, but the

[8] Further information is available on the ILAC website: https://ilac.org/.

scope of its work is different and it focusses on conformity assessment for management systems, products, and services. The IAF is also an association of accreditation bodies (with 91 members in 2022); its main aim is to facilitate international trade by promoting worldwide acceptance of certificates of conformity under the principle 'certified once, accepted everywhere'. The IAF develops additional requirements that accreditation bodies must follow for specific markets, adding to or interpreting the relevant ISO/IEC standards.

One simple example of the value of the international network: I worked for an international company that produced reference materials in the United Kingdom, and the work to characterize the materials was accredited by the UK accreditation body. When changes occurred with the company, it was easy to transfer the work to another department in another country, with UK accreditation replaced by accreditation in the other country. From an organizational point of view, it helped improve efficiency with no impact on customers who were still able to meet their quality system and regulatory requirements.

1.5.2 National level

If your organization is accredited or is seeking accreditation, whether you are a scientist or technician, you are very likely to meet representatives from your national accreditation body, usually a specialist in quality assurance as lead auditor supported by a technical expert, who may be an experienced metrologist from the NMI. Your work may be reviewed, your written records and procedures pored over by an expert in the field, and you could even have someone watching you while you carry out your work. You may also find that experienced colleagues from other departments will do the same. All this is part of the process of achieving and maintaining accreditation, formal recognition that your organization meets a quality standard.

National accreditation bodies are appointed by the government to assess and accredit organizations against national and international standards. In the United Kingdom, the story of the UK Accreditation Service (UKAS) is closely linked to the science of metrology; the work was carried out by a department of the National Physical Laboratory until the formation of UKAS as an independent organization in 1995.

According to UKAS, accreditation brings many advantages to an organization. The public face of accreditation is permission to use the appropriate logo on the website or in publications, and a listing in the UKAS database which can attract customers. Accreditation demonstrates competence and impartiality, free of financial and political pressures, gives confidence to customers, and reduces risks to the organization; accreditation is often an entry requirement to the market for some products and services. A further benefit is that accreditation demonstrates that the organization takes its work seriously; it can show that the management is not being negligent. It is not surprising that in 2022 there were 3000 accredited organizations in the United Kingdom, and accreditation has proved equally valuable to organizations in other countries.

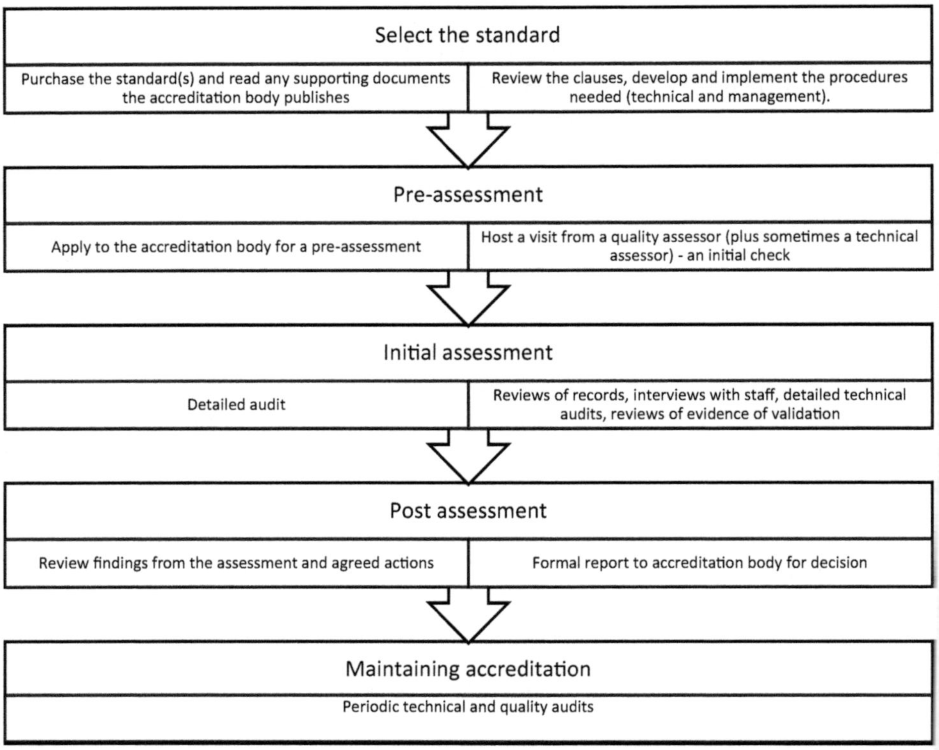

Figure 1.6. A typical process to gain accreditation to a quality management system standard such as ISO 17025.

The process of gaining and maintaining accreditation involves everyone in the organization. A summary of a typical process[9] is shown in figure 1.6.

In the United Kingdom, accreditation to some standards including ISO 17025 is covered by the national accreditation body itself. Accreditation to other quality management system standards such as ISO 9001 can be carried out by other organizations—professional accreditation companies that are themselves accredited by UKAS. Similar to the concept of 'traceability' for metrology, there has to be an unbroken chain of accreditations that leads back to the international system represented by ILAC or IAF.

1.6 Enforcement—legal metrology

The system of accreditation and the use of ISO-type documentary standards is purely voluntary; organizations participate in the scheme because it is of benefit to them. There are cases where quality must be enforced—a simple example being purchasing food by weight, where the temptation for an unscrupulous shopkeeper to steal from customers by fixing the scales could be strong. In the real world, there has

[9] More details are available on the websites of accreditation bodies such as UKAS: https://www.ukas.com/.

to be a fourth building block in the quality infrastructure, a legal framework to enforce quality including traceability to the international measurement system. Almost all countries have established laws and regulations on this issue, referred to as 'legal metrology'.

The starting point for legal metrology was the 1875 Metre Convention. The need for formal cooperation on regulations related to weights and measures was discussed for many years at the General Conference on Weights and Measures, leading eventually to the formation of the International Organization of Legal Metrology (OIML) in 1955 [13]. During the 1970s, more and more states opted to join the OIML and participate in technical conferences. The globalisation of trade during the 1980s was a further impetus to the OIML's work, and the OIML has a close link with the World Trade Organization.

Similar to other international bodies, the OIML is a member-driven organization supported by a small secretariat. By 2022, the OIML had 124 member states, represented by staff from the country's ministry or department responsible for legal metrology.

Today, the OIML develops and maintains standards that are intended to be used by member states as models for their regulations, the aim being to harmonize regulations worldwide. Although there is cooperation with ISO on developing standards (for example, standards on measuring vehicle exhaust emissions), the difference is that the OIML standards are led by governments and ISO standards are largely driven by business and other stakeholders.

We may not be aware of legal metrology standards, but the OIML standards impact every aspect of our daily lives. The standards are intended to protect individuals and organizations, to ensure public health and safety, and to enable trade, at the same time as protecting national interests. They cover measurement issues of public interest including:

- Trade and commerce (from consumers to wholesale to cross-border)
- Services (such as taximeters)
- Utilities (water, electricity, gas)
- Safety (such as road safety)
- Environmental pollution (such as vehicle exhaust, organic pollutants in water)
- Healthcare (such as thermometers)

Standards are drafted by technical committees of staff from member states. Representatives from business can participate as experts in the field or can contribute to discussions through national mirror committees. As the documents underpin laws and regulations, legal metrology has its own carefully defined approaches and vocabulary. The laws and regulations based on the standards usually specify the legal units of measurement and set out requirements on measuring instruments including their accuracy, and they define how the instruments are controlled and the sanctions for breaking the law.

Another difference between ISO/IEC standards and OIML standards is therefore the specification and control of measuring instruments. It is left up to the user of an

ISO/IEC standard to demonstrate that an instrument meets the requirements, and the text leaves the user with a lot of freedom. An OIML standard is more prescriptive, defining the type of measuring instrument, an approved type (model or family), a detailed specification (including the accuracy needed, durability, range, and sensitivity), and requirements for verifying that the instrument meets the specification (before use, periodically, and following repairs). All of these conditions have to be met before an instrument can be offered for sale.

All quality systems address some or all of these issues, but legal metrology requires that you have evidence that would stand up in court. There are other examples of standards with legal status: pharmaceutical regulations require companies to comply with a standard referred to as cGMP (current Good Manufacturing Practice), and medicine regulatory authorities can stop sales of any pharmaceuticals produced in a facility that does not meet this standard.

In addition to developing standards, the OIML promotes mutual recognition of legal metrology systems in different countries, raises awareness of legal metrology, and runs knowledge transfer activities.

1.6.1 Legal metrology at the national level

Three issues have to be considered at national level—policy making, enforcement, and expertise. The aims are to ensure there is a unified national system for measurements and there is evidence to support this, to reduce fraud, and to reduce transaction costs. All of this contributes to the well-being of society.

How this is carried out varies widely from country to country. Some countries use OIML standards for national law; others develop their own based on the OIML standards. Enforcement includes identifying contraventions, applying sanctions, checking instrumentation, and surveillance of what is happening. These enforcement activities may be carried out by inspectors employed by local authorities at the town or district level, or inspectors may be employed by a national service or subcontracted to a commercial organization.

As an example, in the United Kingdom at the time of writing, responsibility for regulations for legal metrology rests with the Office for Product Safety and Standards (OPSS) which is part of the Department for Business Energy and Industrial Strategy. The OPSS leads UK government policy on legal metrology and works closely with the BSI and UKAS. Technical expertise largely rests with the NMI, but some countries have specialist weights and measures laboratories.

Responsibility for enforcing the regulations on weights and measures falls to the local authorities who employ weights and measures inspectors whose job is to verify that weighing and measuring equipment is initially accurate, to visit businesses to check the accuracy of instruments, to check products in shops, and to investigate complaints. Determining whether instruments conform to the standards is carried out by commercial services. In addition to developing the United Kingdom's policy, the OPSS publishes guidance for local authorities, manufacturers, importers, and distributors.

These formal structures for legal metrology are only part of the story of how metrology is embedded in laws and regulations; many regulations refer to measurement in one way or another. The new science of nuclear forensics (see case study 1.4) is a powerful example of the link between quality systems, metrology, and the legal profession.

Case study 1.4. Nuclear forensics—where quality meets politics.
The International Atomic Energy Agency (IAEA) has been concerned about the risks of the illicit use of radioactive materials for many years. In 1995, the IAEA established an Incident and Trafficking Database [14]; since then, more than 3500 cases have been reported of which about 10% have been confirmed as related to trafficking or malicious use.

Interest in scientific research for nuclear forensics has increased in recent years; the applications of the science are shown below, one aim being to enable security services to identify the origin and trafficking routes of any nuclear material seized.

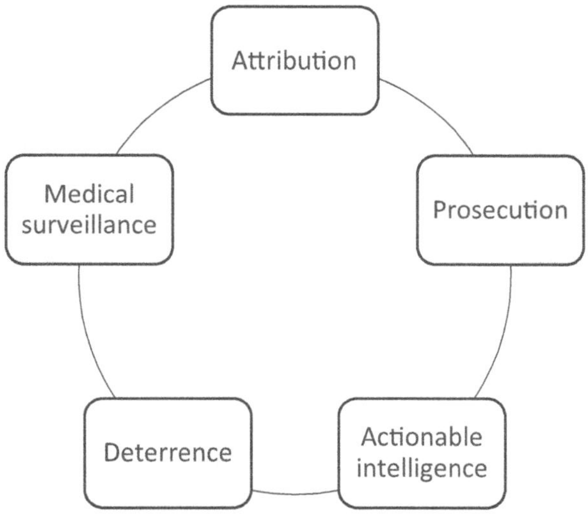

The applications of nuclear forensics
As well as enabling security services to attribute the responsibility for production of the nuclear material seized, nuclear forensics supports the prosecution and identification of the people concerned, deters use of the material, and provides information for treatment of anyone affected.

The science involves determining many characteristics of the material, including its chemical composition, appearance, and structure as well as the radionuclides present. For example, a measurement of the ratio of the activity of different radionuclides can be used to find out if, when, and how the material has been purified, so it can be traced to a particular nuclear plant. Most manmade radioactive materials are subject to regulatory control and thus controlled by states, so the potential diplomatic and political consequences of this work are immense.

For example, there are large stocks of the manmade radionuclide ^{237}Np worldwide [15]. Specialist reference materials are being developed for this radionuclide so that laboratories can validate their measurement techniques with confidence. Neptunium is particularly challenging to attribute, as the processes producing it in the civil nuclear fuel cycle are the same as for its production for military use. The identification of the origin of materials containing neptunium therefore additionally relies on accurate measurements of trace elements (radioactive or otherwise) and other parameters.

The responsibility placed on laboratory quality systems in this field is significant. Traceability of measurements to high-accuracy primary measurement standards, the equivalence of those standards internationally, detailed estimates of measurement uncertainties, and recognised measurement techniques must all be established. The results from measurements may be used in a court, so the laboratory must be able to defend its work under cross-examination; a court ruling in the United States in 1993 [16] established five factors that a trial judge must consider in determining whether a scientific methodology is valid:

1) Whether the technique has been tested
2) Whether it has been peer-reviewed
3) Its known or potential error rate (false positives or negatives)
4) The existence of standards
5) Its acceptance in the scientific community

All of these issues are addressed in quality systems such as ISO 17025, but, due to the potential political impact of the work and the high accuracies needed, nuclear forensics is one of the most challenging examples of implementing a quality system.

Simon Jerome
Fakultet for miljøvitenskap og naturforvaltning, Norges miljø- og biovitenskapelige universitet, Norway

1.7 An example of the quality infrastructure in practice: surface contamination monitors

To conclude this chapter, this example brings together all the different strands of a quality infrastructure.

Anyone who works with radioactivity, from the nuclear industry to nuclear medicine, will be familiar with instruments called 'surface contamination monitors'. These handheld devices are used to detect radioactive contamination on surfaces; they are used to check the workplace to prevent workers from picking up contamination on their hands or clothes and accidentally ingesting the radioactive material. They measure the number of alpha or beta particles leaving the surface per second; the device beeps or shows a count rate on a display, and any contamination found can be cleaned up safely. The workforce in these settings take the instruments for granted; workers just grab the nearest device when needed, assume that it is fit for purpose, switch it on, and carry out the measurements.

In fact, when workers grab the instrument, they are using something that draws on all the different aspects of the quality infrastructure, including scientific, industrial, and legal metrology. The starting point is the realization of primary standards of radioactivity and surface emission at national measurement institutes. There are about 20 national metrology laboratories working in this field, and the KCDB maintained by the BIPM lists all those with peer-reviewed capability. The national laboratories have participated in international comparison exercises for radioactivity and surface emission, also peer-reviewed and published in the KCDB, to show that the national standards are equivalent.

Documentary standards then come into play. Responsibility for these standards falls to an ISO committee (to be precise, ISO Technical Committee 85 (Nuclear energy, nuclear technologies, and radiological protection), Subcommittee 2 (Radiological protection), Working Group 17 (Radioactivity Measurements)). This working group meets once or twice a year and has technical experts appointed by national standards bodies. The working group maintains ISO 8769 which describes the construction and characterization of reference materials for calibrating surface contamination monitors. Manufacturers rely on this standard for the production of reference materials (typically a thin aluminium foil with radioactivity embedded on the surface, mounted on a backing plate); customers accept the reference standards because they comply with ISO 8769. Instrument manufacturers also rely on documentary standards for designing the instrument; IEC 60325 sets out the general characteristics of surface contamination monitors and describes test conditions, radiation characteristics, electrical safety, and environmental characteristics.

The reference sources are used by instrument manufacturers and by calibration laboratories to calibrate the surface contamination monitors to ensure that measurements are traceable to the international measurement system and to carry out other tests to check that the monitors are fit for purpose. The procedures followed are described in another set of ISO standards (ISO 7503); in the United Kingdom, the ISO standard is supported by a Good Practice Guide (GPG14) published by the National Physical Laboratory. The Good Practice Guide was developed by experts in radiological protection and metrology in consultation with stakeholders from the nuclear industry, instrument manufacturers, calibration laboratories, hospitals, and academia.

The calibration laboratories are accredited by UKAS to ISO/IEC 17025 (General requirements for the competence of testing and calibration laboratories) to demonstrate their competence to their customers. As part of their accreditation, the laboratories are expected to follow international or national guidance including ISO 7503 or GPG14.

The final part of the jigsaw is the regulatory regime to ensure that employers are protecting their workforce. In the United Kingdom, this part is covered by the Ionizing Radiations Regulations and the associated Code of Practice, which includes a requirement for surface contamination monitors to be tested annually, the service provided by the calibration laboratories. Enforcement is through the health and safety authorities rather than the weights and measures inspectors, but the point is the same; there is a regulatory requirement behind the metrology.

This is just one example of all the aspects of the quality infrastructure coming together to benefit society. It means that a person in the nuclear industry or the hospital can simply grab the nearest surface contamination monitor and use it, without worrying about whether it is fit for purpose. This pattern is repeated in other sectors for a multitude of purposes, often taken for granted and hidden from view.

References

[1] Bosshardt W and Lopus J S 2013 Business in the Middle Ages: what was the role of guilds? *Soc. Educ.* **77** 64–7

[2] Taylor F W 2003 (1911) *The Principles of Scientific Management* (New York: Dover)

[3] British Library W Edwards Deming (https://bl.uk/people/w-edwards-deming)

[4] BIPM 2019 *The International System of Units (SI)* (Sèvres, France: International Bureau of Weights and Measures) 9th edn (https://www.bipm.org/en/publications/si-brochure)

[5] Milton M J T and Fellag-Ariouet C F 2021 The metric system, the Metre Convention and the BIPM *New Frontiers for Metrology: From Biology and Chemistry to Quantum and Data Science* ed M J T Milton, D S Wiersma, C J Williams and M Sega (Amsterdam: IOS Press) Proceedings of the International School of Physics 'Enrico Fermi' 206

[6] OECDBIPM 2020 *International Regulatory Co-operation and International Organisations: The Case of the International Bureau of Weights and Measures (BIPM)* Organisation for Economic Co-operation and Development (OECD) and International Bureau of Weights and Measures (BIPM) (https://www.oecd.org/regreform/regulatory-policy/events-publications.htm)

[7] Stock M, Davis R, de Mirandés E and Milton M J T 2019 The revision of the SI—the result of three decades of progress in metrology *Metrologia* **56** 022001

[8] BIPM Joint Committee for Guides in Metrology (JCGM) 2012 *International Vocabulary of Metrology—Basic and General Concepts and Associated Terms (VIM)* 3rd edn (Sèvres, France: International Bureau of Weights and Measures) JCGM 200 (https://www.bipm.org/en/committees/jc/jcgm/publications)

[9] CIPM 1999, revised 2003 *Mutual Recognition of National Measurement Standards and of Calibration and Measurement Certificates Issued by National Metrology Institutes* International Committee for Weights and Measures (CIPM) (https://www.bipm.org/en/cipm-mra/cipm-mra-documents)

[10] Pommé S and Keightley J 2015 Determination of a reference value and its uncertainty through a power-moderated mean *Metrologia* **52** S200

[11] Coulon R and Judge S M 2021 An evolutionary algorithm for consensus building in inter-laboratory comparisons *Metrologia* **58** 065007

[12] BIPM Joint Committee for Guides in Metrology (JCGM) 1995, revised 2008 *Evaluation of Measurement Data—Guide to the Expression of Uncertainty in Measurement (GUM)* (Sèvres, France: International Bureau of Weights and Measures) JCGM 100 (https://www.bipm.org/en/committees/jc/jcgm/publications)

[13] OECDOIML 2016 *International Regulatory Co-operation and International Organisations: The Case of the International Organization of Legal Metrology (OIML)* Organisation for Economic Co-operation and Development (OECD) and International Organization of Legal Metrology (OIML)

[14] International Atomic Energy Agency (IAEA) Incident and Trafficking Database (ITDB) (https://iaea.org/resources/databases/itdb)

[15] Jerome S M, Carney K, Essex R, Fassbender M E, Goldberg S, Kinlaw M, LaMont S P, Mackney D, Morrison J J and Nortier F M 2017 Reference materials for neptunium determination *Appl. Radiat. Isot.* **126** 44–8

[16] Cornell Law School Legal Information Institute 2023 Daubert Standard (https://law.cornell.edu/wex/daubert_standard)

Chapter 2

Quality in practice

2.1 Introduction

It is no exaggeration to say that ensuring the quality of your work saves lives. The use of ionizing radiation in cancer therapy is just one such example: an error in the radiation dose delivered to a tumour can result in an underdose (and the tumour is not destroyed) or an overdose (and the patient dies from damage to other organs). Case study 2.1 has other examples of when science has failed, publicly and expensively.

Case study 2.1. When quality fails: cautionary tales.

Sellafield fined after incorrect disposal of radioactive waste

In 2013, the Sellafield nuclear site was fined £700 000 for sending bags of radioactive waste to a normal waste site instead of a special facility. The cause was an error in setting up a waste monitor—in quality system terminology, a noncompliance with validation of a measurement system.

Source: The Guardian, 7 February 2013

Errors in reporting COVID test results at a private laboratory in Wolverhampton, United Kingdom

An error in setting the pass/fail criteria for COVID tests led to some 40 000 false negative test results, resulting in more people being infected and an estimated 23 unnecessary deaths. Again, a validation failure.

Source: UK Health Security Agency report 29 November 2022

Cancer patients overdosed during radiotherapy in Canada during the 1980s

The first medical linear accelerator (LINAC) for cancer therapy to use computer control was introduced during the 1980s. A risk assessment was conducted but the software was not considered (or validated), and software design errors were later found to be a major factor in some patients being overdosed. This incident, amongst others, led to the development of new regulations and guidance.

Source: Report to the British Institute of Radiology, 2012

doi:10.1088/978-0-7503-5107-2ch2

There are millions of measurements carried out every day for trade, healthcare, food safety, materials testing for aircraft, forensics, environmental protection etc. Much can depend on the outcome: criminal sanctions, compensation claims, and the future of a business. Customers, whoever they may be, expect to be able to trust scientists.

One of our roles as scientists is to minimize the risk of errors happening. Some industries, such as the pharmaceutical sector, have strict regulations that must be followed when manufacturing and testing products[1]; other sectors refer to quality assurance management systems such as those described in ISO 17025 [1] or ISO 9001 [2].

Whichever quality management system your organization uses, a quality system covers both scientific and management aspects of your work. The main scientific elements are shown in figure 2.1.

Standing back for a moment, these elements are just common sense, what any organization wanting to reduce the risk to its business would do and how any professional scientist would behave. The general principles are that you need to check that any measurements you are carrying out are correct, you need to check that the measurements are still correct, you need to estimate how wrong the measurements could be, and you have to write down what you are doing. It is also possible to see how following a quality management system also protects you, making sure that you have the equipment and resources to carry out your work.

In a recent report, the US National Academies summarised what is needed: 'All researchers should include a clear, specific, and complete description of how the reported results were reached. […] Reports should include details appropriate for the type of research, including a clear description of all methods, instruments, materials, procedures, measurements, and other variables involved in the study; a clear description of the analysis of data and decisions for exclusion of some data or

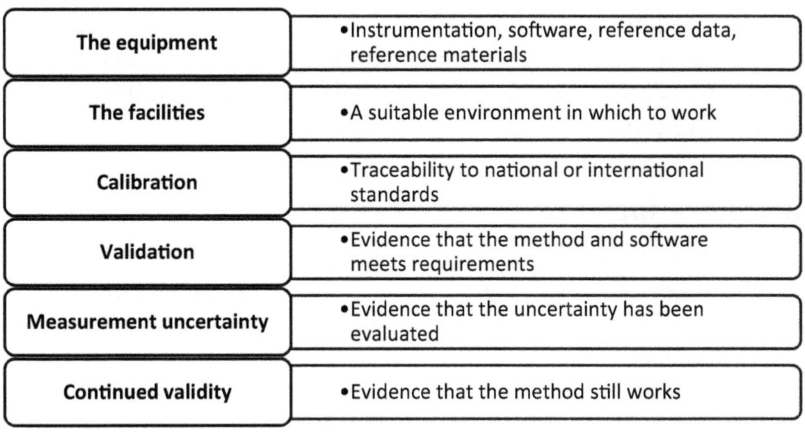

Figure 2.1. The scientific elements of a typical quality management system.

[1] For example, the US regulations are available online: https://www.fda.gov/drugs/pharmaceutical-quality-resources/current-good-manufacturing-practice-cgmp-regulations.

inclusion of other; […] and a discussion of the uncertainty of the measurements, results, and inferences' [3, Recommendation 6-1].

It may seem that using a quality management system is a heavy burden, just additional unnecessary work. A counterargument is that all the work described will have to be done one way or another; the quality system just gives you a framework to use, a systematic approach that will save you time and effort in the long term. This argument is borne out by a study published in *Nature* [4]; concerned by the so-called 'reproducibility crisis' in which results from pure research were sometimes not replicated, some 1500 scientists were surveyed to identify what could be done. Validating measurements, checking results, robust experiment design, closer supervision, and using statistical methods were all listed as approaches that would help improve reproducibility—all of these are aspects of a quality system. This issue will be discussed in more detail in chapter 4.

This section will look at these scientific aspects of the elements of a typical quality system in more detail—the work you are likely to be involved in day-to-day. Chapter 3 covers the management aspects of a typical quality system.

2.2 Instrumentation

One of the technical requirements of a quality system is that your organization commits to providing the right instrumentation and all the associated items such as software, reference materials, and reference data. There are cost implications to this, of course; the quality system can help you make the case to replace obsolete equipment.

The instrumentation must be kept under control and protected from adjustments by restricting access to the laboratory and the instrument settings. A record is needed of the make, model, and serial numbers of the instrument and its location. The instrument may need regular calibration; this should be recorded along with any maintenance carried out and details of any damage, malfunction, or modification. It should be possible to identify easily any instrumentation that is out of service, with a clear label or by segregating it. Many organizations have software packages or databases to make it easy to keep all of these records.

2.3 Laboratory facilities

It may seem obvious that the laboratory itself must provide an environment that will not adversely affect the results. In practice, the building and its environmental controls can be the most expensive aspect of the work, and consequently there can be pressure to economise.

Following a quality system implies that the most important issue is ensuring that the results of the work are fit for purpose, and environmental factors, such as temperature, temperature fluctuations, humidity, dust, vibration, noise, the quality of the electrical power supply, and the impact of other work going on the vicinity, must not be allowed to impact the work. Any important environmental parameters have to be monitored and documented.

2.4 Calibration

The purpose of calibration is to ensure that measurements are traceable to national or international standards. Calibration sets a reference point; this reference point captures the international measurement system in your measurements. All the readings on the instrument or results from the method give the ratio of the measurand to this reference point, either directly from the display or calculated by the software.

There are three ways to calibrate an instrument to ensure traceability. First, the work can be performed by a calibration laboratory that is accredited to a quality standard such as ISO 17025 or by a national metrology institute or designated institute that offers the service. Accredited calibration laboratories are normally listed on the local accreditation body's website, and metrology institutes publish their Calibration and Measurement Capabilities (CMCs) on the Key Comparison Database hosted on the website of the International Bureau of Weights and Measures (BIPM).

Second, the calibration can be carried out in-house, using a certificated reference material or a calibrated transfer instrument (a high-accuracy instrument that has itself been calibrated). A reference material must be supplied by a laboratory accredited to ISO 17025 or ISO 17034 [5] or by a metrology institute with a published CMC. The third approach is to calibrate the instrument using a primary standard—this is only normally possible within a metrology institute.

Part of the evidence that the calibration is traceable to national or international standards is a certificate from the calibration service or the reference material supplier. The certificate should show the logo of the accreditation service and state how their work is itself traceable to a national or international standard. It is good practice to label instruments clearly with the date of calibration (and who carried out the work) and the date of next calibration, so it is easy for users to check that the calibration is valid.

There are cases where no calibration service or certificated reference material is available. A reference material that is accepted by the consensus in the community may be used instead; alternatively, some measurements have to follow a procedure that is defined in a regulation or documentary standard, and the result is accepted as being the correct answer without establishing a link to the international measurement system.

Instruments should be calibrated, then recalibrated, following a documented timetable to maintain traceability, confirm measurement uncertainties, and check the validity of results between calibrations. The International Laboratory Accreditation Cooperation (ILAC) and the International Organization of Legal Metrology (OIML) have published guidelines for determining intervals between calibration [6]. To begin with, the calibration interval should take into account the manufacturer's recommendations, the type of instrument (mechanical wear, electronic drift), the environment (temperature, humidity changes, noise, vibration, etc), the measurement uncertainty required, and data about similar instruments. The interval can also depend on the impact if the calibration were to drift significantly.

Over time, enough evidence will be gathered to decide whether to increase or decrease the interval. The ILAC and OIML guidelines describe several approaches to reviewing the interval between calibrations. The first method is a 'staircase'—if the calibration does not change within a permissible error between calibrations, the interval can be increased. Control charts (see section 2.9.2) can be used to estimate the drift and hence when the instrument should be recalibrated. Some instruments wear or drift depending on how much they are used, and usage can sometimes be a better indicator than time. Another approach is to check the calibration using a high-accuracy transfer instrument and recalibrate if there is a significant difference. In practice, many organizations impose fixed recalibration intervals to meet a regulatory requirement or for simplicity.

All instruments that affect the final result must be calibrated, including ancillary equipment such as balances and pipettes if they contribute to the final result and instruments used to measure parameters that have an indirect effect (i.e. parameters that appear in correction factors to the result). Calibration can be costly and there may be some room for interpretation of this requirement. For example, should a thermometer to measure room temperature be calibrated if the temperature does not affect the final result? It will depend on the circumstances—if the measurement has very low uncertainties, the temperature could have some, as yet unknown, effect and the thermometer should be calibrated. However, a temperature change might have a negligible impact on a less accurate measurement; the effect is hidden in other components of uncertainty, so the purpose of measuring the temperature is just to confirm that the laboratory heating is working: in this case, no calibration is needed.

2.5 Validation and measurement uncertainty

There are two important concepts concerned with checking the quality of your scientific work: validation and measurement uncertainties. The two concepts are treated separately in quality standards, but in practice they are closely related and involve the same work. The main difference is the terminology used to communicate with stakeholders; you are likely to carry out a formal validation study to communicate your work to a regulatory body, and to estimate measurement uncertainties to communicate with customers and an accreditation body. Some fields (chemistry and biology) are more familiar with the term validation, but there is a trend towards measurement uncertainty being the key concept.

The important message from this section is the value of standing back for a moment from what you are trying to achieve and spending a few minutes thinking it through, to make sure that every aspect is under control.

2.5.1 What is validation?

Validation is the process of providing evidence that the measurements you are carrying out produce consistent results and meet the specification—it means showing that the equipment works correctly in your actual setup. That is the formal definition, but my experience is that the main benefit of validation is that it forces you to learn about the instrumentation and the methods. Adopting a structured approach to validation also enables you to tackle measurement systems you are not familiar with; you learn as you go along. Most importantly, validating a measurement system means that you should not have that horrible moment in the middle of the night when you wake up thinking 'was that result really correct?'.

Validation is essential in industry, but it is also valuable in fundamental research. Going through a validation process captures issues you may need to include in subsequent scientific papers. Thinking through the details in advance can avoid wasted experimental studies—I worked on a measurement of the rest mass of the neutrino where the pre-work showed that a proposed method, promising on the face of it, could not work as the measurement uncertainties were too large (this work was, of course, never published). There can be nasty surprises hidden away in instrument software—I remember a software package which made mistakes in corrections for radioactive decay, a simple calculation that I assumed was straightforward—it is always worth checking.

The clauses in ISO 17025 and other quality standards do not describe validation in detail. There are many excellent sector-specific guides (such in analytical chemistry [7] and pharmaceuticals [8]); the aim of this section is to raise awareness of the issue and give some generic practical advice. Of course, this advice should always be used in the context of regulatory requirements and company or institute procedures.

The extent of validation needed depends on the application and can be difficult to determine. It challenges assumptions about the work: are you measuring what you think you are, is it accurate, does the method work in your laboratory, do staff have the skills and experience to use the method? Professional judgement is needed to decide how much effort to invest, as cost has to be balanced against risk. The effort needed will depend on the demands of the project or service.

Different scientific fields use different terminology—'method performance characteristics', 'method performance', 'performance properties'—but the purpose is the same, and the following sections aim to make this clear by the context. The validation process described below covers 'single-laboratory validation': it tests the method as implemented in your laboratory. It does not look for the effect of different implementations of the method in different laboratories—this is covered by participation in multilaboratory studies or proficiency test exercises (see section 2.8).

Although validation can be time consuming, this is nothing in comparison with the problems that can arise if something goes wrong or the time you could waste troubleshooting. Carrying out validation studies also helps you devise and write the operating procedures, as you will know any limitations. With careful planning, the work is not burdensome. The next sections are a step-by-step guide to conducting a validation study.

2.5.2 The validation master plan

The first step is to decide which instruments or methods need validating, as carrying out all the tests can be time consuming and in some cases the risks are low. Every case is different, and you will need to exercise your technical judgement.

How much effort to put in to validation depends on the technique you are using and the application. Detailed validation is needed if it is required by regulations or by customers or if a new method developed in-house is being used. Less effort is needed if the method is a modification of a previously validated method, if the method is set out in documentary standards or in good practice guides, or if it is very similar to other validated methods. The work may be minimal if the instrumentation has already been validated by the manufacturer (clinical instrumentation sometimes falls into this category) or if it is a revalidation after changing equipment or staff (a benchmarking study may suffice, in which a test artefact is measured before and after the change).

A technique developed in-house will therefore need more detailed validation than using a published, tried, and tested approach. Even with these in-house techniques, experience and records of similar approaches can be used to design and steer the steer the validation studies and to decide their extent. The interpretation of 'detailed', 'less detailed', and 'limited' is an issue for professional judgement—omitting tests that are not relevant, using simple or complex statistical tests, and reducing the number of replicate measurements can help to reduce the amount of work needed.

A comprehensive list of the equipment and the steps used in the method will be needed so you can make an informed decision. One important issue to consider is whether the readings affect the results directly or indirectly. For example, the temperature and humidity in the laboratory may only affect results indirectly, so validation studies can be very simple.

Armed with this list, it is possible to devise a document called a Validation Master Plan. This is not an explicit requirement of some quality systems, but a Validation Master Plan is useful and it is something that you may encounter. The aims of the Validation Master Plan are to set out the organization's policy for validation, including which standards or best practice guides will be followed; to list the instruments and processes that should be validated; and to say who is responsible for what. The Validation Master Plan should also describe the process to be used for validation (such as the V-model, described below).

The Validation Master Plan may include instructions on the format and contents of documents to avoid the need for you to devise your own—these documents may be shared with regulators, auditors, and sometimes with customers. The procedure to follow if changes are needed may also be described, along with the policy for revalidation (when the work should be repeated).

2.5.3 The V-model—a process for a validation study

A common process used for validation is the so-called V-model—I first used this model when working in the pharmaceutical sector, where it is in common use, although its origins go back to the aerospace industry in the 1980s. I found the model to be helpful particularly when setting up a new laboratory or developing a new service, as it can be used as a simple project plan, it helps show progress to colleagues, and it saves time because it avoids writing lengthy reports at the end. Using a recognized approach such as the V-model also helps communication, as colleagues are familiar with the terminology. It is scalable—you can use the approach for everything from validating a single instrument to setting up a new laboratory. Another powerful benefit is that it can be used even if you are not familiar with the technique.

The V-model (figure 2.2) is a step-by-step process, so it is not ideal for projects that evolve such as software development, for which other process have been devised (software validation is discussed in more detail below). However, my experience is that it works well for most scientific studies. The approach is shown in the diagram below; the terminology is from the pharmaceutical sector, but you may come across other jargon—it is the principle that is important. The clever part of the V-model is that each stage in the specification arm of the model is mirrored by tests to ensure that reality matches expectations. You may not need to split the work down into these exact steps (it depends on the organization and regulations), but they can help guide your thinking.

The V-model has two phases: specifying and purchasing the equipment and software, and carrying out the validation studies. In practice, it is rarely as linear as this, and the work goes back and forth between phases.

Phase 1: Specifying and purchasing the equipment and software

Step 1: What does your customer want? The User Requirements Specification

The User Requirements Specification is an outward-looking document that records what you want to achieve from the user's point of view. This means looking only at the outputs and ignoring the method, instrumentation, and software—it means discussions with the customer to avoid specifying something that is not possible. The User Requirements Specification is sometimes referred to as the 'Measurement Quality Objective' [9].

The questions to be answered in the document depend on what you are aiming to achieve but will typically include:

- The purpose of the measurement system—what exactly is it intended to measure, over what range, to what accuracy, and to what sensitivity?
- What regulations, standards, or customer requirements must the system meet?
- What is the throughput needed—how long will it take to carry out each measurement?
- How will results be communicated to the customer?

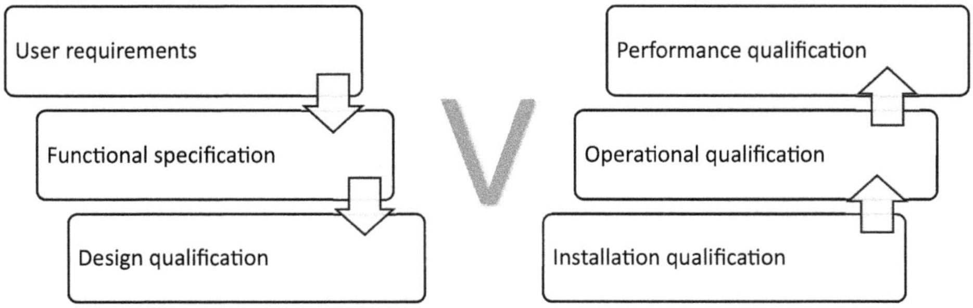

Figure 2.2. The V-model of method validation.

A simple example of the contents of a User Requirements Specification for a system to measure radioactivity in soil samples is shown in figure 2.3. In text form, the aim is 'to determine the ^{60}Co content in Bq g^{-1} of a sample of soil, over a range of 10–1000 Bq g^{-1}, with an uncertainty better than 10% at the 95% level of confidence over the range. The density of the soil samples can be from 2.0 to 4.0 g cm^{-3}. The samples may also contain ^{137}Cs at an activity of 1000 Bq g^{-1}'.

The advantage of defining a User Requirements Specification is that it is performance based. The laboratory should be free to choose any method that meets customer requirements (there are exceptions to this; sometimes the method is set in regulations).

Step 2: What method shall we use? The Functional Specification

The Functional Specification is inward-looking; it describes the technique and instrumentation needed for the organization to meet the User Requirements Specification. Although it sets out the technical details, it stops short of listing specific models of instruments.

Starting from the User Requirements Specification, the Functional Specification considers relevant documentary standards, best practice guides, and the scientific literature and sets out in some detail the method that will be used. Some justification for the choice may be included; this will be useful if the approach is challenged by the management or the customer, or if the work is part of an academic study and may be published.

Item	**Example**
Measurand	Activity (Bq) of ^{60}Co in soil samples
Range	Quantification range 10-1000 Bq g^{-1}
Uncertainty	Better than 10% at the 95% level of confidence over the range
Matrix	Soil of density 2.0 – 4.0 g cm^{-3}
Sensitivity	Detection limit 1.0 Bq g^{-1}
Possible interferences	^{137}Cs at an activity of up to 1000 Bq g^{-1}
Documentary standards	ISO20042 ISO11929
Throughput needed	10 samples per day
Reporting	The following information must be included in a measurement certificate: - Sample number - Description of sample - Date of receipt - Results - Etc

Figure 2.3. Typical contents of a User Requirements Specification.

If the User Requirements Specification is performance based, the Functional Specification can be creative. It may be possible to combine different techniques so that the method will be more efficient; for example, analysis of samples for alpha-emitting radionuclides can be sped up using a simple technique to eliminate samples free of such radionuclides rather than subjecting every sample to a full analysis.

The Functional Specification may include more details on the required performance of the technique, with more focus on implementation than on customer requirements. As well as describing the technical specification of the instrumentation needed, the Functional Specification covers issues such as the utilities that will be needed (electrical supplies, compressed air, gases, water, vacuum, etc) and any environmental control systems needed (temperature, humidity, vibration, noise, etc). Any ancillary items that will be needed (balances, pipettes, thermometers, calibration standards, glassware, etc) should also be listed; it is all too easy to forget something essential.

The Functional Specification also takes account of any constraints. It should consider people factors such as the operability of the method (how much training and expertise are needed). Other issues to consider include the cost, the time needed, regulatory permissions, waste disposal, and safety problems that affect how the technique will be implemented.

Finally, a list of the documents (operating manuals, parts lists, engineering diagrams, certificates of conformity, calibration certificates, etc) that will be needed is useful, along with any training requirements.

In the previous example, a Functional Specification might have the contents shown in figure 2.4.

The Functional Specification feeds into the purchasing procedure if you are buying new equipment. The technical specification for the equipment can be lifted out and used as part of the invitation to tender process.

Step 3: What equipment and facilities will we use? The Design Qualification

The Functional Qualification has set out the generic requirements without specifying the exact equipment to be set up or purchased. The Design Qualification is similar to the Functional Specification but moves into the details of what will be implemented, and typically lists equipment chosen including the manufacturer, model, and the engineering diagrams of any items to be constructed. The manufacturer's specification should be compared to the requirements set out in the Functional Specification and a justification included for any deviation from the Functional Specification. The laboratory facilities available should also be compared to the requirements—the environmental control systems available and utilities needed.

The Design Qualification is a 'parts list'—a detailed list of all the instruments, equipment, laboratory consumables, calibration artefacts, and documents (manuals, calibration certificates, engineering diagrams, etc).

Phase 2: The validation studies

The first three steps ensure you have clear and detailed specifications for the work. The next three steps cover the scientific aspects of the work. An efficient way to record the results from the studies is to use the approach from the pharmaceutical sector by writing the reports in advance; each report is written in the form of a workbook or a checklist, with a short procedure set out for each of the tests and

Item	Contents of Functional Specification
Measurand	Activity (Bq) of ^{60}Co in soil samples: the method (high resolution gamma spectrometry) to be used and the justification from the scientific literature including traceability to national or international standards.
Range	Quantification range 10-1000 Bq g^{-1}: generic details of the instrument, ancillary equipment and laboratory consumables needed to meet this requirement, with justification.
Uncertainty	Better than 10% at the 95% level of confidence over the range: calculations or evidence to show that this uncertainty can be met.
Matrix	Soil of density 2.0 – 4.0 g cm^{-3}: calculations or evidence to show that the method will work for these samples.
Sensitivity	Detection limit 1.0 Bq g^{-1}: calculations to show that this level of sensitivity can be met or evidence from the scientific literature.
Possible interferences	^{137}Cs at an activity of up to 1000 Bq g^{-1}: justification that the method is specific to ^{60}Co in the presence of ^{137}Cs.
Documentary standards	Technique-specific standards (in this case, ISO20042, ISO11929) and an explanation of how the proposed method meets these standards
Throughput needed	10 samples per day: a calculation that shows how this requirement is met.
Environmental conditions / facilities	Laboratory conditions (20-25 °C): details of the laboratory infrastructure needed, the power supply, arrangements for earthing equipment, the laboratory layout, anti-vibration mounts for balances, supply of liquid nitrogen etc. How the temperature in the laboratory will be monitored and controlled. High background radiation from sources stored in vicinity: how the spectrometer will be shielded from interfering radiations.
Operability	To be operated by laboratory technicians: a description of the training courses needed and requirements for the software (inputs, outputs, interface).
Reporting	The format for results to be communicated to the customers and data storage: a detailed description of the format for the results and how raw data will be controlled. A list of the supporting documentation needed (operating manuals, engineering diagrams etc).

Figure 2.4. Typical contents of a Functional Specification.

space left for the results. The blank workbook is then reviewed by a colleague and may also be approved by the quality manager or the section head.

The results from each of the tests is added to the workbook, and the completed document goes out for peer review and approval again. This may seem a heavy process, but the advantage is that each step is carefully peer-reviewed twice. In reality, this process can save time as there is no need to write a lengthy report at the end and, because you are adding results as you go along, you can quickly identify any problems or missing data.

Best practice is to define pass/fail criteria for each of the tests, using statistical analysis where possible. A summary of useful statistical tests is given in figure 2.5, and the website of the US National Institute of Standards and Technology has a valuable resource [10]; such tests are easy to carry out using software packages or spreadsheets, although advice from a statistician is always valuable. The tests should

Useful statistical tests for validation studies: hypothesis testing

Hypothesis testing refers to a set of statistical techniques that can be used to compare results in an objective way. It is useful for validation studies as it gives a way to demonstrate that results are unchanged if input parameters are changed – for example, you may want to demonstrate to an auditor that changes in the laboratory temperature do not impact the results.

The steps in hypothesis testing are:

- Define the 'null hypothesis' - usually 'the results are unchanged'. The 'alternative hypothesis' is that the results are changed significantly.
- Define what you mean by 'significant' – the level of statistical significance, the so-called p-value. This is the probability you will make the mistake of concluding that the results are different when in fact the null hypothesis is correct. A value of p=0.05 is usual.
- Carry out the measurements – for example, one set of measurements at one laboratory temperature and a second set at a different temperature.
- Calculate the statistic, compare the value to tabulated values for the p-value you have chosen for the number of degrees of freedom (which depends on the number of measurements in the set). If the statistic is higher than the tabulated value, the conclusion is that the results from the two sets of measurements are different and you have to consider corrective actions.

Comparing means of two data sets: Student's t-test

The t-test can be used to compare the means of results from two data sets, for example, to compare the results using two different instruments. Assuming a normal distribution and the data sets are independent and have the same variance, a two-tailed non-paired t-test can be used.

The t-statistic is given by the difference between the means divided by the standard error of the difference:

$$t = \frac{m_1 - m_2}{\sqrt{\frac{s_1^2}{n_1} + \frac{s_2^2}{n_2}}}$$

where m_1 = mean of dataset 1, $s_1^2 = \frac{\sum_1^{n_1}(x_i - m_1)^2}{n_1 - 1}$ and n_1 = number of data points in set 1,

m_2 = mean of dataset 2, $s_2^2 = \frac{\sum_1^{n_2}(x_i - m_2)^2}{n_2 - 1}$ and n_2 = number of data points in set 2,

and the number of degrees of freedom is $n_1 + n_2 - 2$.

If the value of t is greater than the tabulated value, the conclusion is that there is a difference between the two data sets. For a validation study, this means investigating the reason for the discrepancy and working out how to solve the problem, or possibly increasing the measurement uncertainty.

Comparing the variance of two data sets: F-test

An F-test can be used to check whether two data sets have the same variance. For example, you may want to compare the results from measurements of similar items – the mean result would be expected to be different, but you may expect no statistical difference between variances as the same technique is being used.

The F-statistic is given by:

$F = \frac{s_1^2}{s_S^2}$, choosing the labels 1 and 2 so that $F \geq 1$.

Similar to the t-test, the value of F can be compared to tabulated values for the F-statistic for different significance levels for $(n_1 - 1)$ and $(n_2 - 1)$ degrees of freedom. If the value of F is greater than the tabulated value, it can be concluded that the two data sets have different variances. For validation studies, this test may be useful to compare results obtained by two different operators; if the variances are discrepant, it means that the method may not be robust and the procedure may have to be changed.

Figure 2.5. Statistical tests useful for validation studies.

Testing the goodness of fit: linear regression and the chi-squared test

A chi-squared test is useful to confirm the linearity of a method. Typically, one might expect the output (y) and the input (x) to be a first-degree polynomial of the form:

$$y = ax + b$$

Linear regression determines the values of the slope (a) and intercept (b) to minimise the sum of the squared residuals:

$$\sum_{i=1}^{n} (y_i - (ax_i + b))^2$$

The chi-squared statistic is given by:

$$\sum_{i=1}^{n} \frac{(y_i - (ax_i + b))^2}{ax_i + b}$$

The value of the chi-squared statistic can then be compared to tabulated values for the required significance for $n - 1$ degrees of freedom. If the value of chi-squared is greater than the value tabulated, the test is failed and the response of the system (the output) is not a simple straight-line function of the input.

It is not uncommon in validation studies for this test to be failed. First, it does not consider the different uncertainties on the individual data points – weighted least-squares fitting should be used. Second, higher-order terms may be needed in the polynomial to fit the data accurately; however, more terms should be added with caution and the plot examined carefully as a good fit to the data points may not result in a curve that is accurate between the data points.

Comparing several data sets: ANOVA

The analysis of variance (ANOVA) refers to an array of different methods to compare several sets of data. It is useful if you wish to compare results from several different sets of measurements, to test the robustness of the method.

For validation studies, a 'one-way ANOVA test' can be used to test if there is a discrepancy between data sets. The test involves calculating the 'within set' sum of squares and the 'between set' sum of squares. For example, if there are *k* data sets:

Data set	Mean	Data points	Number of data points
1	m_1	x_{1i}	n_1
2	m_2	x_{2i}	n_1
..k	m_k	x_{ki}	n_k
Overall mean	M		

The 'within sets' sum of squares SS(within) is given by: $\sum_{i=1}^{k} \sum_{j=1}^{n_j} \left(x_{ij} - m_i\right)^2$

The 'between sets' sum of squares SS(between) is given by: $\sum_{i=1}^{k} n_i (m_i - M)^2$

The ratio $\frac{SS\ (between)}{SS\ (within)}$ follows an F-distribution with $(k - 1), (\sum_{i=1}^{k} (n_i - 1))$ degrees of freedom.

This test just shows whether or not the data sets are discrepant. If it is failed, further statistical tests may be needed to identify where the problem lies.

Figure 2.5. (Continued.)

match across to the specification so that there is a documented link at all the stages in the validation process from the specification to demonstrating that the specification has been met. Some of the tests will fail, as perhaps the specification was too tight; this does not mean that the validation study has failed, but space should be left in the workbook to record any actions arising (for example, you may find that there are some constraints you should include in the operating procedure). Failing a validation test is not the end of the world, but you must document the consequences.

All of the documentation should be thorough, with a clear record of the procedures followed, identification of who carried out the work, and references to any supporting documents and records.

The workbook style is not the same as a normal scientific or laboratory report, but this doesn't matter as it will only be used by colleagues or by auditors and it will contain all the information you need for a report or paper.

Step 4: Have we got the equipment and facilities we need? The Installation Qualification
Anyone who has worked in a laboratory in academia or in industry will be familiar with the hours of frustration hunting down the missing instruction manual or the mains adapter for the instrument you want to use. The Installation Qualification helps to reduce this wasted effort and is worth every minute of writing. It also helps deal with the usual chaos that surrounds the delivery of new equipment, avoids frantic calls to the supplier a few days later for missing items, and reminds you of checks to carry out on delivery.

The Installation Qualification is based on the list of equipment and facilities from the Design Qualification and confirms that they have been delivered or are in place. It would normally include:

- A record of the equipment—manufacturer, model, serial number, its location, and its inventory number
- A list of any parts delivered
- Confirmation that the equipment has been delivered undamaged—a visual inspection and, if possible, confirmation that the instrument at least switches on
- Documentation such as:
    - Engineering drawings and the as-built drawings
    - Manuals
    - Calibration certificates
    - Certificates of conformity
- Confirmation that the laboratory environmental controls and utilities are in place and operating correctly

This step can be very useful; it is easy to forget to ask the supplier for something important in the excitement of the delivery of a brand-new instrument. The as-built drawings are easy to forget and are essential for the long term. The basic tests confirming that instruments are undamaged and operate may be sufficient to confirm that the supplier can be paid.

An example of a typical entry in an Installation Qualification document is given in figure 2.6.

Item	Notes
Spectrometer	Manufacturer: *XYZ* Model: *AB* Serial number: *123456* Location: *Laboratory A* Visual inspection: *No damage observed* Operates: *Confirmed that the instrument powers up and software runs with no errors reported* Operating Manual: *Stored in cupboard Y* Certificate of conformity: *No 123456 stored in cupboard Y* Check carried out by: *J Smith* Date: xx/yy/zzzz

Figure 2.6. An example of an entry in an Installation Qualification document.

Step 5: Do we meet the Functional Specification? The Operational Qualification

The Operational Qualification is the bulk of the scientific work and covers how you demonstrate that the measurement system meets the specification. It covers the whole process, from start to finish, but may involve validating some aspects of the process in more detail.

The tests to carry out depend on what is being validated; sector-specific guidance can be found in documentary standards and in good practice guides. This section sets out the main issues to consider, but every case will be different. The related topic of software validation is discussed below. The outputs from this stage are a document recording the results of a series of tests, a detailed uncertainty budget, and a document that describes how to operate the equipment and run the process (an 'operating procedure').

A useful place to start is a series of 'scoping studies', sometimes called 'phase 1 validation'. This means setting up all the equipment, learning how to run the method, and carrying out an initial calibration. The scoping studies help you refine your plans for validation but are not a substitute for the detailed work.

A typical Operational Qualification records the evidence that the method measures what we are assuming it does and that it works consistently to a stated uncertainty over the range needed. Tests will typically need to cover the factors shown in figure 2.7, but not all the tests are relevant in all cases. With careful design, it is possible to combine tests to avoid repeating work.

1) Identity
    1. This test confirms that the method measures what you are expecting it to measure. It is useful if the aim is to determine the content of a chemical compound etc in a sample—a typical approach is to use an artefact or a reference material which has been characterized using a different method or certificated by an independent laboratory. This test is usually only carried out once.
2) Selectivity
    1. The aim is to demonstrate unequivocally that the method works in the presence of anything that could interfere with the measurement. Examples include other compounds in a sample that could confuse the measurement system—the test involves adding (spiking) samples with the compounds that could interfere and checking that the method

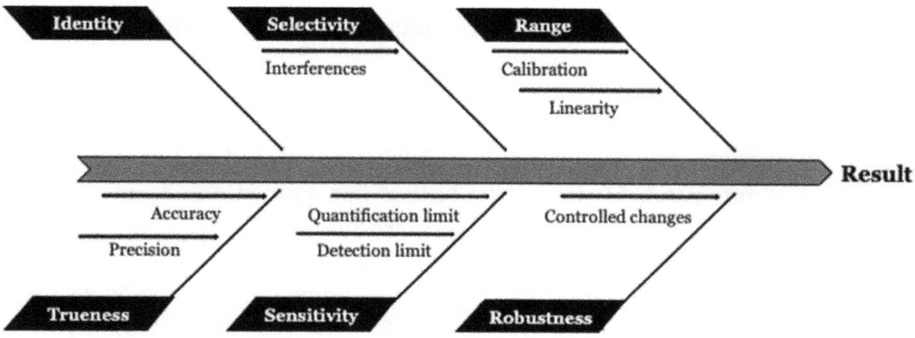

Figure 2.7. The typical contents of an Operational Qualification.

discriminates against them. It can be made quantitative by calculating the ratio between the response to the target compound and the interfering compound (the selectivity index).

2. This does not just apply to chemical analyses though. There can be many different interferences—electronic noise from other equipment in the area, background radiation from sources stored in the same laboratory. Another approach to test selectivity is to use a different method, with a different response to interferences, and compare the results.
3. The more sensitive and sophisticated the measurement is, the more vulnerable it may be to unexpected interferences. There are even examples in astronomy, including unexplained signals in a radio telescope being traced eventually to a microwave oven in the kitchen[2].

3) Range, linearity, and calibration
   1. The aim of this test is to provide evidence that the method works over the whole range specified for the measurand with the uncertainty needed. It is related to the calculation of 'trueness' and the estimation of uncertainty.
   2. The test starts with calibrating the instrument and/or the method—ensuring that measurements are linked to the international measurement system and traceable to national or international standards. This means using a reference material (certificated by an accredited supplier) or a calibrated instrument (such as a certificated current source to calibrate an ammeter) or arranging calibration of the instrument by an accredited service.
   3. Calibration sets a reference point. All the readings on the instrument or results from the method give the ratio of the measurand to this reference point, either directly from the display or calculated by the software. The next challenge is to extend this calibration over the whole range, to

[2] See https://www.theguardian.com/science/2015/may/05/microwave-oven-caused-mystery-signal-plaguing-radio-telescope-for-17-years (accessed 23 January 2023).

show that the reading or results are correct within the required measurement uncertainty over the expected range of the measurand.

4. One approach to testing the linearity is to measure reference materials of known content starting with a material with 80% of the minimum expected value and finishing with a material that has 120% of the maximum value (varying the output in a known way from a reference instrument has the same effect). Carrying out the test below and above the extremes of the range improves the uncertainty in the calibration at the top and bottom of the range. The 'range' may also be defined as the lowest quantifiable measurement up to the maximum possible.
5. Typically, at least six data points are needed to check linearity, each data point in duplicate or triplicate and measured in a random order. The response (output) is then plotted against the known content (input). Analysis depends on the rigour needed for the validation; it can be as simple as fitting a straight line to the results by linear regression and inspecting the residuals to check that no point deviates by more than a given percentage from the straight line. If more mathematical rigour is needed, goodness of fit tests such as a chi-squared tests may be needed, and weighted linear least squares fitting may be required. 'Linearity' in this context can be a misnomer, as the response may not be linear over the whole range. Higher-order terms may be needed in the calibration function to improve the goodness of fit. The uncertainty in the calibration will feed into the calculation of 'trueness' and measurement uncertainty, and may depend on the value of the measurand—these can be calculated from the fitting algorithm.
6. Linearity therefore can cover two issues, the linearity of response of the instrument and the linearity of the entire process, with the latter being the key issue. It is possible to adjust the process so that all the measurements fall into the linear range of the instrument. The output from this part of validation is evidence that the technique works over the full range of the measurand within a given uncertainty.

4) Trueness

1. 'Trueness' is a term that is used to describe the closeness of a result to the accepted reference, the 'conventional true value'. Measurement uncertainty captures the same issues but uses different terminology.
2. Whatever terminology is used, there are two factors to consider. The first factor is that repeated measurements will not return exactly the same result, due to small uncontrolled variations in parameters such as electronic noise that affect the measurement—this may be called the precision, the random error, or the Type A uncertainty.
3. Precision may be broken down into:
   – The *repeatability* of a measurement: the spread in results (standard error) for measurements carried out over a short time period by one operator.

–*Intermediate precision*, sometimes called the *reproducibility*: the spread in results (standard deviation) for measurements carried out over a longer time period by different operators.

4. Confusingly, the term reproducibility is also sometimes used to refer to the difference between different laboratories carrying out the same method.
5. The precision is quantified by calculating the standard deviation of repeated measurements—typically five to ten replicates. It may vary over the range, so measurements may have to be repeated at different levels.
6. The precision can be improved by taking the result from a measurement to be the mean of replicate measurements, in which case the spread is given by the standard error of the mean (SEOM = st dev/$\sqrt{n}$ where n is the number of replicates). The dependence on $\sqrt{n}$ means that, after a few replicates, the benefit of using more may be small. It can also be time consuming to carry out a large number of replicates; other longer-term factors such as instrumental drift come into play, and the standard deviation may increase.
7. The second factor affecting 'trueness' is that measurements can be different from the conventional true value, due to intrinsic issues with the method and how it has been implemented in the laboratory. The difference is not reduced by taking repeated measurements. This factor may be called the accuracy, the bias, or the systematic error, and it is captured in the calculation of Type B uncertainty.
8. Bias may be broken down further into:

   - *Run bias*—the difference between the result obtained in one run of measurements and other runs
   - *Laboratory bias*—the difference between the result that the laboratory obtains and the conventional true value, due to how the method has been implemented
   - *Method bias*—the difference between the result and the conventional true value, inherent to the method itself. The method bias may be zero by definition if a particular method is required by regulations.

9. Bias can be quantified by comparing results from measurements to those obtained using other methods or using reference materials or instruments. A laboratory working in isolation cannot distinguish between laboratory bias and method bias; comparisons with other laboratories using the same method are needed to determine the laboratory bias.

5) Sensitivity

1. 'Sensitivity' concerns determining the lowest value of the measurand that can be detected above zero. If the measurand is close to this limit, the measurement uncertainty is likely to be high, and it may indicate that another method should be used. For some applications, such as

determining the content of pollutants in environmental samples or impurities in pharmaceuticals, sensitivity may be the most important part of validation.

2. Determining the sensitivity can be complex. Assumptions about the method can break down at low levels of the measurand; for example, at low concentration levels, interactions with the surfaces of materials can become important for chemical analyses. Another example is that the assumption of a normal distribution may not be correct at low levels. Some fields, where sensitivity is a key parameter, have adopted Bayesian statistics as a more rigorous approach for calculating the sensitivity (this is outside the scope of this book, and one can refer to the many sector-specific scientific papers and guidance documents). It is a big topic and, as one colleague once said to me, 'never has so much been written about so little'.
3. However, there are some relatively simple approaches that are in common use. The first approach is semiquantitative; a blank test sample is prepared or a reference instrument is used, and replicate measurements are made, typically ten or more. A known quantity of the measurand is added or the setting is increased, and the measurements are repeated. The process is repeated, recording the number of times that the result is greater than the blank sample, and the sensitivity is determined by choosing the level of the measurand that is detected with an acceptable probability of detection. This method requires a large number of measurements and has high uncertainties (see figure 2.8).
4. An alternative is to calculate the detection limit based on statistics. The simplest technique is to take multiple measurements of a blank sample,

Concentration	Detected	Not detected
0	0	10
0.1	1	9
0.2	3	7
0.3	6	4
0.4	7	3
0.5	10	0

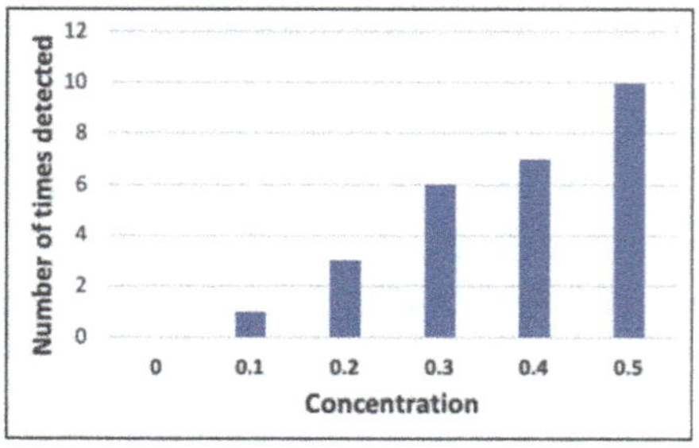

Figure 2.8. Semiquantitative method to measure the sensitivity. If the measurand is the concentration of a chemical in a sample, the sensitivity can be estimated by carrying out replicate measurements (ten or more) on a blank sample, adding the chemical to the blank and recording how many times the chemical is detected by the method. In this example, a concentration of 0.3 would be detected roughly 50% of the time. It is semiquantitative as there are high uncertainties unless a very large number of replicate measurements is carried out.

calculate the standard deviation of the results, and estimate the detection limit as 3.3 times the standard deviation. Another more rigorous technique is to consider the risks of the errors that can occur at low levels of a measurand:
   - A false positive can occur when the measurand is reported as present when it is, in fact, not present.
   - A false negative can occur when the measurand is reported as absent when it is, in fact, present.
5. Assuming a normal probability distribution (which may not be correct), a 'critical value' (l_c) can be defined as a multiple of the standard deviation of replicate measurements of a blank sample (or zero input), so $l_c = f \times$ st dev. If the result from the measurement is greater than l_c, the measurand is reported as being present. The factor f depends on the risk that can be accepted for reporting a false positive—a probability of $\alpha = 5\%$ is commonly used, corresponding to $f = 1.65$.
6. This does not address the risk of reporting false negatives though—if the true value is close to the critical value, the probability of reporting that the measurand is absent is high (50% just by chance). A second quantity, the detection limit (l_D), is defined where $l_D = f' \times$ st dev. The factor f' depends on the risk that can be accepted for reporting a false negative—for a probability of $\alpha = 5\%$, $f' = 3.3$.
7. In summary, if the result (x) from a measurement is greater than the critical value (l_c), the measurand is report as present at a value of x. If the result from a measurement is less than the critical value, the measurand is reported as not detected with probability of $\alpha\%$ with a detection limit of l_D. The definition of the detection limit can lead to confusion, as positive results can be reported that are below the detection limit.
8. A 'quantification limit' may also be defined. The quantification limit is the lowest value of the measurand that can be determined reliably (it is sometimes used to the set the lower limit of the range). It is typically set at two or three times the detection limit, but it may also depend on customer requirements or by calculating when the measurement uncertainty reaches an acceptable level.

6) Ruggedness
1. Also called 'robustness', ruggedness refers to confirming that the method works if it is subjected to small controlled changes in parameters that impact the results. Issues such as the laboratory temperature or humidity, changes in suppliers of laboratory consumables, changes in the time taken for different steps—even the technique used to mix solutions can have some effect.
2. Quantifying ruggedness means carrying out repeated measurements on samples, artefacts, etc, making deliberate changes to parameters, and using a statistical test such as a t-test or ANOVA to look for significant

differences due to the change. There may be many parameters to consider, but if a published method is used these parameters may already have been considered.

3. The outcomes from this test include constraints to be included in the procedure—strict limits on the laboratory conditions, for example—and information for calculating the measurement uncertainties. Testing ruggedness may even result in changes to the method itself to ensure that the method is reliable.

An example of the contents of a typical Operational Qualification for the example of the measurement of ^{60}Co is given in figure 2.9.

Test	**How**	**Relationship to the Functional Specification**
Identity	Measurement of a certificated reference material of ^{60}Co and confirmation that the method identifies the presence of ^{60}Co	Evidence that the method identifies ^{60}Co
Selectivity	Measurement of reference materials containing a mixture of ^{137}Cs and ^{60}Co and demonstrating that the method identifies ^{60}Co.	Evidence that the method identifies ^{60}Co in the presence of ^{137}Cs.
Range	Calibration using a certificated reference material and measurement of 6 or more reference materials containing ^{60}Co over a range of 80 % of the minimum activity specified to 120 % of the maximum activity specified. Linear regression with a chi-squared test to confirm linearity. Mathematical modelling (with references to the scientific literature) of the corrections needed for different sample densities.	Evidence that the method is linear over the range of activities and sample densities.
Trueness	Replicate measurements of a certificated reference material, and calculation of the standard deviation. Calculation of the measurement uncertainty.	Estimation of one component of the Type A uncertainty for inclusion in the uncertainty budget. Evidence that the measurement uncertainty meets the specification.
Selectivity	Determination of the minimum detectable activity using published formulae (in this case, ISO11929)	Evidence of meeting the detection limit criterion.
Ruggedness	Repeated measurements on different days by different operators, different laboratory conditions, of a certificated reference material. A t-test or ANOVA could be used to compare results.	Evidence of operability and confirmation of repeatability

Figure 2.9. Example of the contents of an Operational Qualification.

Step 6—Does the method work in the real world? The Performance Qualification
The previous step has provided evidence that the method can meet the Functional Specification. It has also resulted in a written procedure that can be followed by suitably trained and experienced staff.

Until this point, validation has mostly used reference materials or well-characterized reference instruments. The performance qualification is where the procedure meets the real world; it is the step that shows that the method meets the User Requirements Specification—it shows that the method does not depend on the expertise of one individual, reducing risk to the organization.

Several colleagues should be trained on the operating procedure and asked to work independently, carrying out the measurements on 'real' samples, artefacts, or instruments. Statistical tests such as t-tests or ANOVA are useful in comparing results. It may take a long time to test the procedure thoroughly, with different environmental conditions and different batches of consumables. There is a balance to strike between the time to carry out the tests, the risk of errors, and opportunities to catch any long-term problems through regular quality control tests or participation in laboratory proficiency test exercises.

In the example, a typical Performance Qualification would include the issues in figure 2.10.

The output from this step is a report comparing the results to the User Requirements Specification, which may be used to refine the operating procedure and amend the uncertainty budget. The work may also help define the frequency of quality control tests.

Test	How	Relationship to the Functional Specification and User Requirements Specification
Ruggedness	Replicate measurements of 'real life' samples by different operators on different days. A t-test or F-test can be used to compare the results.	Evidence that the specification is met for range, trueness (uncertainty), selectivity, identity, sensitivity and robustness
Turnaround times	Records of the time taken for measurement, analysis and reporting.	Evidence that the specification is met for turnaround times

Figure 2.10. Typical contents of a Performance Qualification.

2.5.4 Last step—authorization

The final step in a validation study is authorization, normally by a senior manager or scientist and the quality manager. The results from the study will be reviewed to confirm compliance with the technical specification and to confirm that the validation studies were carried out as required by the quality system.

Actions that arise from authorization may include updating staff training records, labelling equipment to show the next calibration dates, and arranging maintenance and calibration contracts.

All scientific studies require some form of measurement, and validation is the cornerstone of measurement. The importance of validation is shown in the next few pages in a detailed example from a university (case study 2.2).

Case study 2.2. Validation and measurement uncertainty in practice: the approach used by the University of Southampton, United Kingdom.

Introduction

This case study is a detailed example of the value of validation in practice, showing how the guidance can be interpreted and implemented to build confidence in the results from measurements. The example concerns the measurement of plutonium in environmental samples, the aim being to ensure that contaminated materials can be disposed of safely. With the potential growth in the use of nuclear power over the next few decades, reassuring the public that measurements of plutonium are reliable and independent of the nuclear industry is important.

As well as teaching and research, the University of Southampton offers services on a commercial basis, bridging the gap between academia and industry. One such service is GAU-Radioanalytical (GAU), an established radioanalytical and consultancy provider that has been serving the nuclear and wider industry since 1987. GAU specialises in the assessment of radioactive and trace metal contaminants in all areas of the nuclear industry and in the environment and also undertakes fundamental and applied research, instrument and method development, and industry training. GAU is accredited to ISO/IEC 17025:2017 and is committed to delivering innovative analytical solutions whilst maintaining the highest levels of data quality and customer service.

GAU performs a range of analytical techniques that are accredited to ISO 17025. Each method is validated to demonstrate that it is fit for purpose and to determine the performance parameters of the method. One such method is the determination of alpha-emitting plutonium isotopes using alpha spectrometry. The following sections describe how a method validation is performed, using the analysis of alpha-emitting plutonium isotopes in solid matrices by alpha spectrometry as a case study. It is a complex validation process, but it illustrates the value of a planned approach and shows how one set of measurements can be used to address several aspects of validation, reducing the work needed.

More details on the method validation process that we use at the university may be found in Magnusson and Örnemark [7].

Scoping the method

As part of the validation process, the scope of the method needs to be defined. The scoping exercise should outline the procedure that is being validated, identify the key stages that are critical to the performance of the analysis, define the analytes and samples that will be analysed, and identify any interferences that may impact on the measurement.

In this example, the procedure describes the determination of alpha-emitting plutonium isotopes ($^{239+240}Pu$ and ^{238}Pu) in solid environmental samples (for example, in soil, sediment, concrete, and biota). It's not possible to measure the alpha particles emitted from the sample directly, as they are absorbed within the sample; chemical separation procedures have to be followed to extract the plutonium from the sample using chromatography followed by alpha spectrometric measurement of the purified fraction. To measure how well the separation procedures have worked, a radioactive isotope of plutonium (^{242}Pu) is added to the sample and its recovery is determined. Plutonium is separated using anion chromatography, and the purified plutonium is electrodeposited onto stainless steel discs; the final measurement of Pu isotopes is performed by alpha spectrometry. The activities of Pu isotopes are measured relative to the ^{242}Pu yield monitor, effectively correcting for chemical recovery and detector efficiency.

To add to the complexity of the analysis, the sample type will determine the initial treatment of the samples (i.e. acid leach or total dissolution). Samples which are likely to contain refractory plutonium (for example, from weapons fallout: soils and riverine sediments) will require borate fusion to ensure complete dissolution. Soluble plutonium, such as that delivered by marine discharge, can be quantitatively leached using aqua regia (for example, marine sediments).

It is not possible to distinguish between ^{239}Pu and ^{240}Pu using alpha spectrometry due to the similar alpha energies of the two nuclides. A combined $^{239+240}Pu$ activity concentration is therefore reported.

On top of these problems, the presence of significant quantities (>50 μg) of some other elements in the sample will result in degradation of the alpha spectrum, making the final measurements more difficult. The presence of other alpha-emitting radionuclides (notably americium and polonium isotopes) will also result in false positive results.

The simple aim of the work (measurement of plutonium isotopes in environmental samples) is therefore much more complex than it may seem at first. Figure 2.11 is an outline of the procedure followed to dissolve and concentrate the sample itself. Figure 2.12 shows how the plutonium is separated. Figure 2.13 shows the end result: thin deposits of plutonium on metal discs (in a form that can now be measured) and the results from alpha-spectrometry measurements.

Accuracy

As you can see, the process is complex with many stages; the challenge is how to gather enough evidence to show that it works reliably, without a workload that becomes burdensome. The first issue we address at GAU is the accuracy of the results.

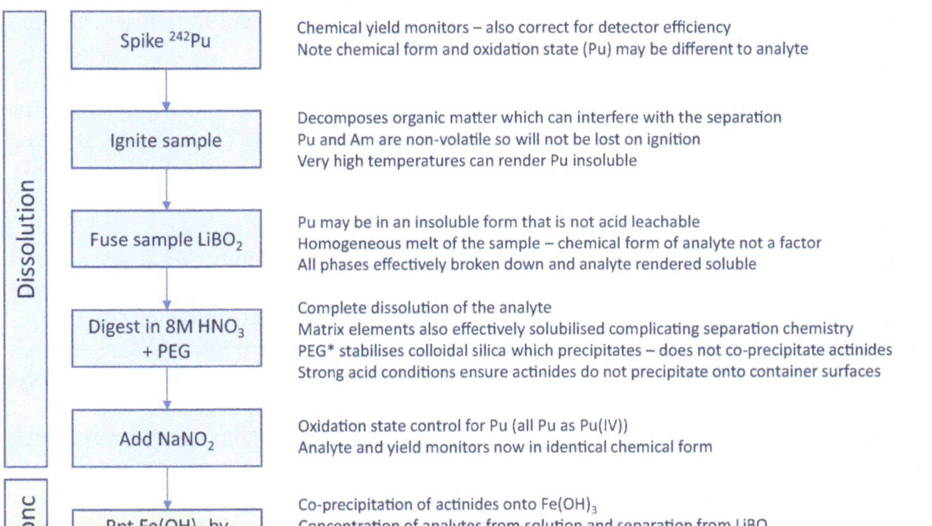

Figure 2.11. Key stages in sample dissolution and preconcentration.

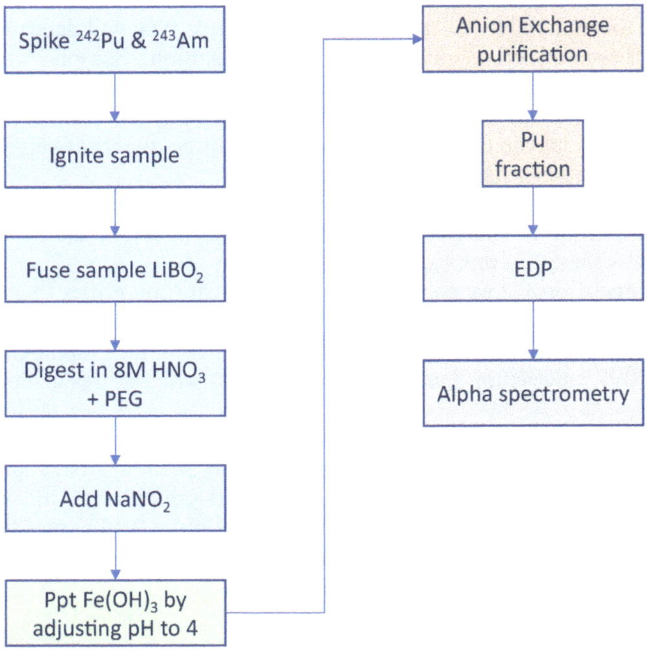

Figure 2.12. Key stages in the analytical separation of plutonium.

Ideally, the accuracy of a measurement is determined through the analysis of certified measurement standards and participation in proficiency testing exercises. In practice, the range of standards is limited and doesn't cover all the different types of sample needed. An alternative method is to add a known quantity of a plutonium isotope to a sample and compare the result from the measurement to the known input (called 'spiking' in radiochemistry jargon).

In this case study, accuracy was assessed using certified reference materials, proficiency test participation, and analysis of spiked samples; the results from the measurements are shown in table 2.1 and plotted in figure 2.14. These results are used in determining the measurement uncertainty (see later in this case study).

Precision

The second issue we address at GAU is the precision of the method: an assessment of the variability in results observed through repeated analysis of a single sample. In this type of measurement, precision depends on many factors that influence the sample preparation and analysis stages. We usually assess precision through replicate analysis of a sample over a long time scale by multiple analysts, to ensure that long-term effects (for example, seasonal variation in temperature) are taken into account.

For ^{239}Pu, we have data from replicate analysis of one particular sample going back 11 years (figure 2.15), showing the value of maintaining long-term records in our quality system (an initial determination of precision would be much quicker than this of course, but having long-term data builds confidence over time). Calculations show that the precision is 13% (95% confidence level).

Robustness

For us, 'a robust process' means a process that is not sensitive to changes in conditions such as reagent purity reagent concentration or pH. For example, for some chemical

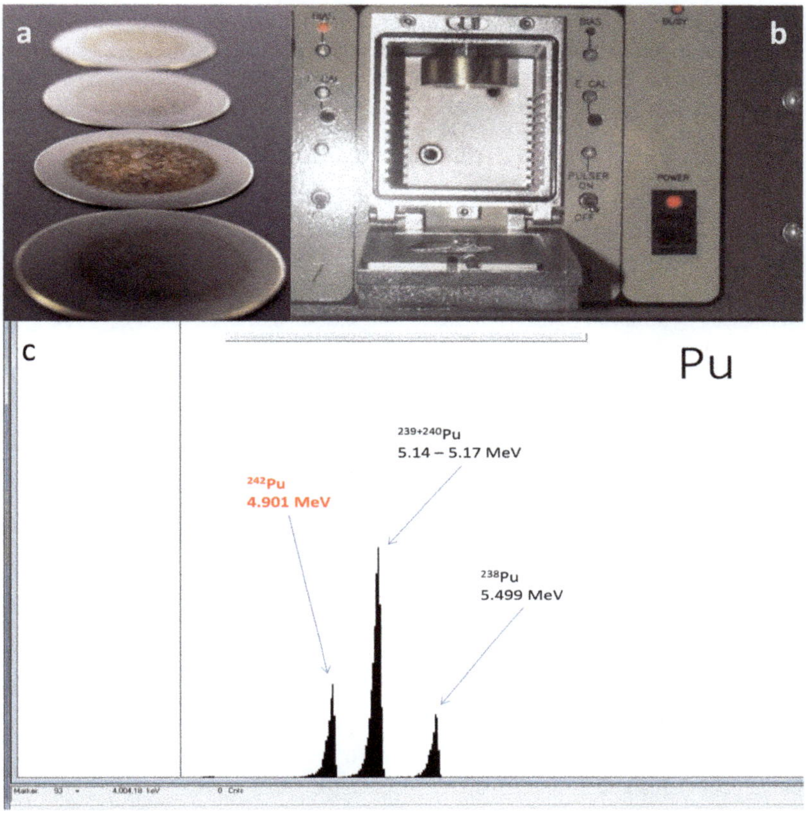

Figure 2.13. The dissolution, concentration, and separation stages result in thin deposits of plutonium on stainless steel discs, as shown in (a). The alpha particles emitted from the disc are measured using a solid-state detector (b). (c) shows a typical alpha spectrum, with peaks corresponding to the different plutonium isotopes in the sample.

separations a broad variation in reagent concentration could be tolerated without impacting on the performance of the stage, whereas for other stages a small change in conditions could have significant impact. By knowing to what degree a stage is impacted by changes in acid conditions, appropriate procedures can be specified to control these conditions where required.

Usually such assessments are made through a combination of literature-based assessment and, where such information is not available, experimental assessment of the sensitivity. For example, plutonium is retained on an anion exchange column from 8 M HNO_3. Under these conditions, Pu(IV) is strongly retained. Changing the acid concentration from 8 to 6 M or up to 12 M is unlikely to significantly impact on the retention of Pu(IV), so strict control of the HNO_3 concentration is not required. However, the failure to add sodium nitrite may result in some of the Pu being present as Pu(III), which is not retained on the anion exchange column, so this is therefore a critical step in the analysis.

Ruggedness

We define 'ruggedness' to mean that the method will work for a wide variety of samples with different chemical compositions, when used by different analysts over a long time

Table 2.1. Results from measurements of different types of sample, showing the difference between the expected plutonium content and the measured values. Three types of sample are used: reference materials provided by the International Atomic Energy Agency (IAEA), samples used in UK national proficiency test exercises, and samples prepared by GAU by spiking.

Test type	Material type	^{239}Pu (measured) [Bq g^{-1}]	+/–	Expected ^{239}Pu activity concentration [Bq/g]	Deviation [%]
Reference material	Sediment (IAEA 135)	0.210	0.014	0.213	−1.4
	Sediment (IAEA 367)	0.042	0.002	0.038	+10.5
	Sediment (IAEA 368)	0.032	0.003	0.031	+3.2
	Sediment (IAEA 384)	0.099	0.007	0.108	− 8.3
	Sediment (NIST 4357)	0.009	0.001	0.0104	−14
Proficiency test	Aqueous	6.4	0.3	6.06	+5.6
	Aqueous	0.020	0.001	0.019 48	+2.7
	Aqueous	15.9	0.9	15.34	+3.7
	Aqueous	10.9	0.6	11.25	−3.1
	Aqueous	16	1	17.29	−7.5
	Aqueous	5.9	0.3	6.138	−3.9
Laboratory spiked matrix	Brick (2 g)	0.77	0.09	0.74	3.70
	$CaSO_4$ (plaster simulation) (2 g)	0.74	0.09	0.74	−0.68
	Glass (2 g)	0.71	0.08	0.74	−3.96
	Sand (2 g)	0.72	0.09	0.74	−2.99
	Wood (4 g)	0.71	0.08	0.74	−4.03
	Ins. foam (0.5 g)	0.76	0.09	0.74	1.91
	Rubber (2 g)	0.77	0.09	0.74	4.28
	Filter paper	0.71	0.08	0.74	−4.35
	Electric cables (3.5 g)	0.68	0.10	0.74	−8.08

period. For us, the difference between robustness and ruggedness is that the former is incorporated in the process itself but the latter is less under direct control. Both are related to accuracy and precision.

A method must therefore be tested using the range of sample types to be analysed and over the mass ranges required to achieve the desired limits of detection. We also have to consider the chemical form; plutonium discharged from nuclear sites is usually in a soluble form and therefore can be solubilised using a digestion with HNO_3/HCl. However, if the plutonium has been heated to high temperatures (for example, during a nuclear weapons test) the plutonium may be present in a highly insoluble form which requires either HF or a fusion-based technique to solubilise it.

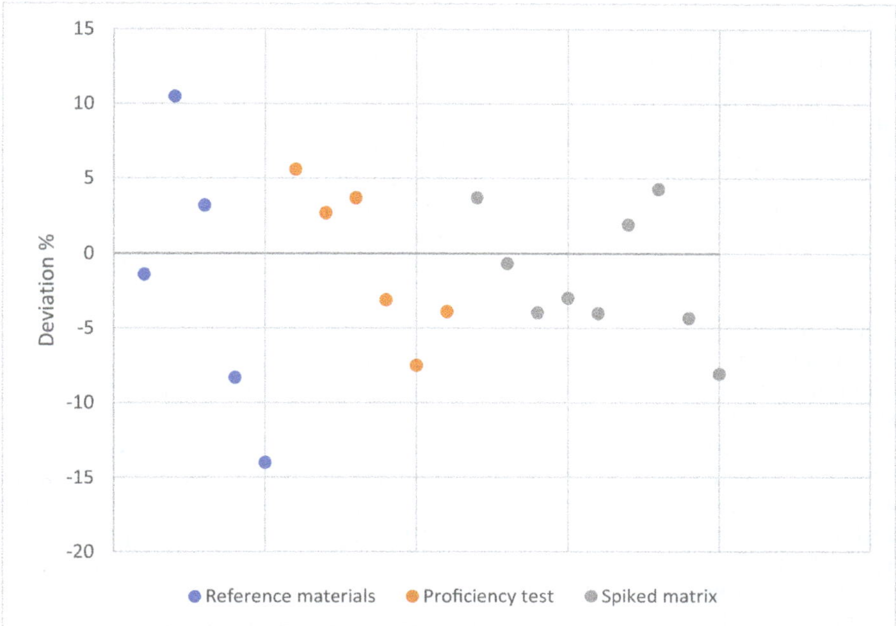

Figure 2.14. Plot of the deviation of the results from the measurements from known values. The scatter in measurements of reference materials is due to the low activity of plutonium in the samples.

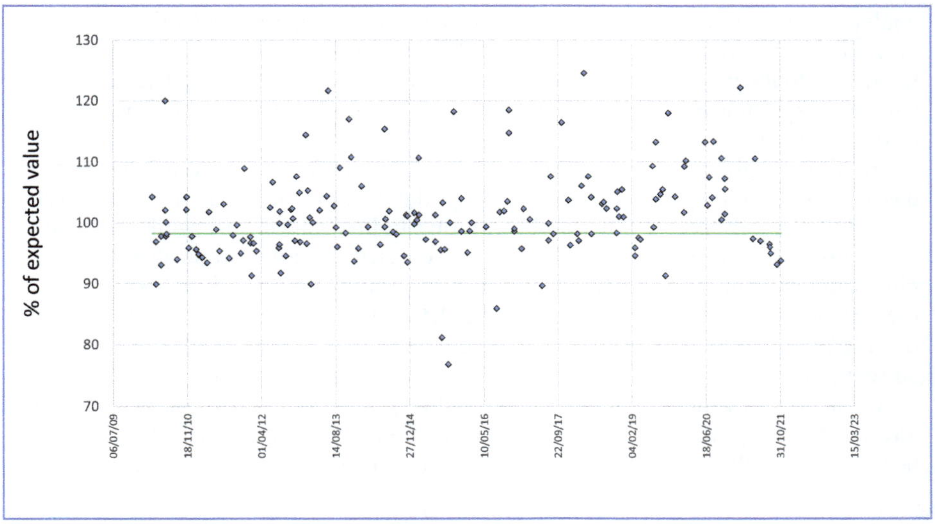

Figure 2.15. Replicate analysis of the ^{239}Pu content of a single sample, to determine the precision.

In general, an analytical procedure should perform effectively independent of the instrumentation or analyst. In practice, small differences in measurement equipment and subtle changes in the way analysts follow the procedure will introduce additional uncertainty. To assess this, the procedure should be undertaken multiple times by different analysts and using different measurement instruments. Such an assessment will identify any additional components to the overall method uncertainty and provide feedback on how sensitive the procedure is to instrument and analyst variability.

Linearity

We do not routinely assess the linearity of the measurement process. This may seem to be contrary to guidance on validation, but the only step in the process that depends on the amount of plutonium in the sample is the final measurement: the plutonium content of samples is so low that the instrument is not challenged.

If needed, we confirm linearity by measuring samples containing different known levels.

Sample integrity

Validation for us includes checking that the sample has not been affected by storage conditions before measurements start; we consider sample storage and treatment as part of the whole process.

In this example, aqueous samples must be acidified to prevent precipitation of plutonium onto the walls of the container resulting in a loss of the analyte from the sample matrix. If the sample already has particulate material present, the sample may need to be filtered. For solid samples, preservation tends to focus on the prevention of degradation of the sample (particularly for biological materials), and this usually involves refrigeration or freezing of the sample.

One method that we use to check the influence of storage conditions is to review the data used for assessing precision.

Selectivity

Selectivity (also called 'specificity') means demonstrating that the analysis detects the analyte of interest and should not be impacted by the presence of other species. In practice, interference from other species will always occur to some extent. It is therefore important to identify which species will, if present, result in a false positive result along with the magnitude of the effect for a given activity of the interferent. This is a measure of the selectivity of the analysis.

In this example, beta- and gamma-emitting radionuclides will not interfere with the measurement as they are not detected by the instrument used in the final step. However, the presence of other alpha-emitting radionuclides with similar energies to the plutonium isotopes of interest could be detected, resulting in an overestimate of plutonium content. The chemical separation procedure has been developed to remove these radionuclides. This will never be completely effective, and a small proportion of the interferent is likely to be present in the final purified fraction, but as long as this can be reduced sufficiently it will not impact on the final measurement.

To determine whether the interference is likely to be significant, the proportion of any interferent initially present that is likely to remain in the final purified fraction must be determined. We use the term 'decontamination factor' (DF) to quantify this effect:

$$\mathrm{DF} = \frac{I_{\mathrm{int}}}{I_{\mathrm{fin}}}$$

where I_{init} is the activity of the interferent in the initial fraction taken for separation and I_{fin} is the activity of the interferent present in the final purified fraction of the analyte.

Measurement of blank samples

There is a risk that some of the reagents used in the process could themselves contain radioactive materials—even spectroscopic-grade reagents do not routinely specify the content of rare radioactive isotopes (this can be an issue for naturally occurring radionuclides such as uranium in particular). There is also the possibility of cross-contamination of samples in the laboratory, despite very careful housekeeping.

We therefore analyse blank samples (samples we know contain no plutonium because they come from areas with no possible contamination). The blank samples are processed in an identical manner to that of normal samples, and the activity concentrations of the analyte are measured in the purified fraction.

If the radionuclide of interest is detected, ideally we would test all reagents and purify them before use if needed. This is significant additional work and is not always feasible, so sometimes, if the analyte concentration in the blanks is found to be consistent, we subtract the value from the results of measurements. In the example of ^{239}Pu, contamination of reagents at significant concentrations is highly unlikely, and only the potential for sample cross-contamination is assessed through routine analysis of procedural blanks.

Limit of detection

The limit of detection of the analytical procedure is calculated based on an internationally recognised methodology and using operationally relevant input data. For plutonium analysis, the limit of detection depends on the measurement time, detector background and counting efficiency, the chemical recovery of the analyte, and the mass of sample taken for analysis. Values for each parameter are selected that are consistent with those typically encountered during the method validation. These values should not differ significantly from those achieved in routine measurement.

Assuming a 5 g sample size, a 100 mBq internal tracer, 60% recovery, a measurement time of 100 000 s, a background count of 10, and a detector efficiency of 30%, the limit of detection achievable using this procedure is 0.0002 Bq g^{-1}.

Method uncertainty

The results from the validation studies feed into our estimates of measurement uncertainty in the results, but we ensure that all possible sources of uncertainty are considered carefully, so we treat this calculation as a separate topic. For this, we follow the guidance on estimating uncertainties described by the European Co-operation for Accreditation [11] (which matches that in the *Guide to the Expression of Uncertainty in Measurement*) expressing the different components as Type A (determined from replicate measurements) or Type B (using other approaches).

In analytical chemistry, it is normal to assess sources of uncertainty arising from the method which are constant for all samples (for example, uncertainties associated with subsampling, weighing, volume dispensing, recovery determination, instrument calibration, nuclear data). We work systematically through the process, identifying all the possible input quantities (see figure 2.16), and then evaluate the effect of each quantity on the final result. We use data obtained during the validation process. In some cases, we refer to the scientific literature or use our own professional judgement in the estimate, based on our experience of similar analyses.

Over time, our understanding of the process improves, and we update the estimate of uncertainties accordingly.

Summary

A summary of the approach we use is given in table 2.2.

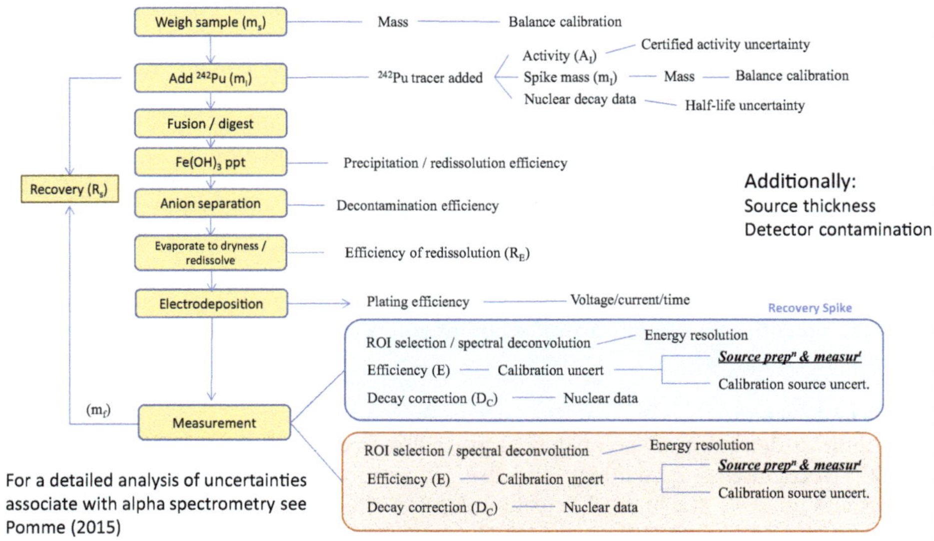

Figure 2.16. A summary of sources of uncertainty in the measurement of ^{239}Pu. We refer to the scientific literature (such as Pommé [12]) to support the calculations.

Table 2.2. A summary of the main issues covered by our validation process.

Parameter	Definition	Method of assessment	Factors impacting performance
Accuracy	How close the measured value is to the true value.	Measurement of reference materials (matrix matched where possible). Determination of the overall bias in the measurement. Proficiency tests.	Sample dissolution efficiency, instrument calibration, yield determination.
Precision	How reproducible the measurement is.	Replicate analysis of a standard. Calculation of uncertainty budget.	Variation in key methodological or environmental factors through the analysis. Instrument stability.
Robustness	Dependence of the performance of the method on changes to procedural conditions.	Variation of key parameters to see how they impact on accuracy and precision of measurement.	pH. Separation flow conditions. Reagent composition.

(*Continued*)

Table 2.2. (*Continued*)

Parameter	Definition	Method of assessment	Factors impacting performance
Ruggedness	Dependence of the performance of the method on the individual or instrumentation undertaking the analysis or changes in sample composition or environmental conditions.	Variation of key parameters to see how they impact on accuracy and precision of measurement.	Operator. Instrument. Sample matrix. Environmental conditions.
Linearity/range	The response of the method to the concentration of the analyte.	Measurement of standards covering the range of concentrations to be measured. Standard concentrations should be selected to extend past the desired measurement range.	Signal suppression. Instrument dead time. Saturation of extractants.
Sample integrity	The stability of the sample.	Assessment of factors that may result in uncontrolled loss of analyte or degradation of the sample.	Storage conditions. Handling procedures. Preparation procedures.
Selectivity/ specificity	How selective the analysis is to the analyte of interest and what species, if present, will interfere with the separation or measurement of the analyte.	Assessment of interferences likely to impact on the analysis. Spiking of samples with selected interferents and determining the impact of the final measurement. Determination of separation factors. Analysis of samples representing the range of matrices to be analysed to check for compositional interferences.	Radiometric interferences (radiometric analysis). Isobaric interferences (mass spectrometry). Matrix interferences (sample composition). Contamination from reagents or equipment.
Reagent/ equipment blanks	Contribution to the analyte from reagents and containers.	Measurement of procedural blanks.	Reagents, containers.
Limit of detection	The minimum activity or concentration of the analyte that could be detected at a specified confidence level.	Measurement of instrument blank and calculation of limit of detection.	Electronic stability. Detector integrity. Environmental factors.

Conclusion
This case study is one example of our work to develop and implement analytical procedures, contributing to the body of knowledge in the field and enabling the university to offer a unique service for environmental protection.

Following the systematic approach described and summarised above has led to independent recognition of the quality of the work through accreditation by the United Kingdom Accreditation Service. We feel that it also supports our students in making the transition from academic life to working in industry, if they decide to follow that path in their careers.

Professor Phil Warwick
University of Southampton, United Kingdom

2.5.5 Measurement uncertainty

Measurement uncertainty is closely related to validation, and the studies needed are often the same. The difference is often a communication issue: regulators in the pharmaceutical sector will look for evidence of validation, whereas an assessor checking conformance to ISO 17025 may focus more on the uncertainty budget.

Before summarizing how measurement uncertainties are estimated, a word or two of reassurance may be needed. First, it is much easier to estimate uncertainties in practice than to read about the theory. Second, the level of detail you will need depends on the application; a national metrology institute will have to study the uncertainties in great detail and mathematical rigour as the reference materials and calibrations will be the basis of perhaps hundreds of thousands of measurements, but most applications will not need that level of detail, and capturing the main components of uncertainty is all that quality standards require.

The concept of traceability described in chapter 1 emphasizes that the results of all measurements are ratios to internationally defined standards. No matter how carefully a measurement is performed, there is always some doubt in the result. Measurement uncertainty is the quantitative description of the doubt that surrounds the result—the range in which the 'true' value is expected to lie. The result from a measurement therefore needs three numbers to describe it—a minimum value, a maximum value, and the level of confidence that the true value lies between the stated minimum and maximum value. This is usually expressed as a value, plus and minus a standard uncertainty at a stated level of confidence.

The recommended approach to estimating measurement uncertainties is set out in a publication referred to as the 'GUM' (*Guide to the Expression of Uncertainty in Measurement*) [13]. Published by the BIPM, the GUM is the definitive and comprehensive international guide; it sets out a systematic approach starting with analysing the measurement process in detail and identifying all the factors that could influence the result. The factors are then combined using the 'law of propagation of uncertainties', the end point being a probability distribution for the result described in terms of a standard uncertainty. There are many introductions to the subject available, such as *The Beginner's Guide to Uncertainty of Measurement* [14].

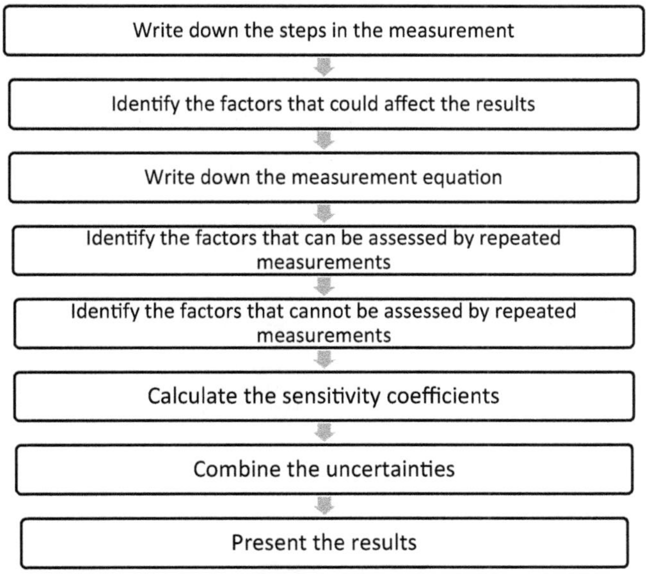

Figure 2.17. The steps in estimating measurement uncertainty.

Not all laboratories have used the GUM in the past, and some historical practices may persist, relying on repeated measurements to determine the spread in results or combining components linearly (this approach was adopted to err on the safe side by overestimating the uncertainty). Apart from its international recognition, there are strong advantages in using the GUM: repeating measurements does not uncover components of uncertainty that are common to all the measurements, the GUM methodology enables results from different laboratories to be compared fairly, it helps understand the underlying science, and it aids communication with auditors.

The steps in estimating the measurement uncertainty are shown in figure 2.17.

The process begins by working systematically through the whole measurement process and identifying the factors that could influence the final result—the different components of measurement uncertainty. Some possibilities are set out in figure 2.18.

A cause–effect diagram may be useful to help identify the different components of uncertainty, check for missing components, and avoid double-counting. For example, one of the main parameters in the measurement equation could be the reading on a voltmeter. The reading will depend on the calibration of the instrument (an imported uncertainty from the calibration certificate) and possibly also on the laboratory temperature and humidity (for which the uncertainty may be unknown). There may be sub-branches as well, as the calibration of the instrument itself could depend on the laboratory conditions at the time of calibration. The main parameters form the main trunk of the diagram with the contributing factors working outward, as shown in figure 2.19; an example of a rigorous analysis is given in case study 2.3.

The measurement equation describes how the final result is calculated from the input quantities and is derived by working through the whole measurement process, so it can be complex. The effect of some of the input quantities (for example, the temperature in the laboratory) may not be known but can be captured in the

Source of uncertainty	Examples of influence factors – the components of uncertainty
The instrument	Resolution of readings, drift, settings
The item being measured	Purity, changes in the item during storage
The laboratory environment	Changes in temperature, humidity, noise, vibration, electrical power fluctuations
The measurement process	Linearity of response of instrument, repeatability and reproducibility, carry-over from previous measurements, changes in background
Imported uncertainties	Uncertainties in calibration of instruments and ancillary equipment, reference data, reference materials
Operator skill	Changes in sample preparation, selection of reagents used, cleanliness of glassware, sample contamination, bias in taking readings

Figure 2.18. Possible sources of measurement uncertainty.

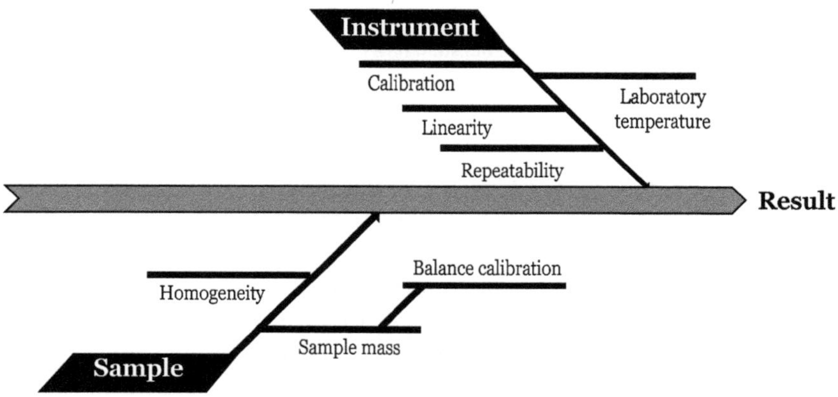

Figure 2.19. A simple example of a cause–effect diagram to identify components of measurement uncertainty (the content of an element in a sample).

measurement equation by multiplying the measurement equation by correction factors to be determined experimentally:

$$\text{Result} = F(a_1, a_2, a_3\ldots) = f(a_1, a_2, a_3\ldots) \times C(a_n) \times C(a_{n+1})\ldots$$

where the parameters 'a' are the input quantities and 'C' are correction factors. This is called a hybrid model as it has both theoretical and empirical components, and it covers most practical cases. The input quantities 'a' may also be functions of other input quantities.

Part 6 of the GUM [16] points out that the choice of measurement equation depends on many issues, including the availability of a theoretical form, the ranges of the output and the input parameters, and the target measurement uncertainty.

The next step is to classify the components of uncertainty in the measurement equation. The GUM defines two types of uncertainty (A and B) depending on how

Case study 2.3. Uncertainties in nuclear magnetic resonance measurements.
Nuclear magnetic resonance (NMR) is a versatile and effective technique for establishing the structure and purity (mass fraction content) of complex organic compounds. An example of its application is the characterization of mycotoxin reference materials. These reference materials are used to validate methods for checking the safety of foodstuffs—the storage of grain and nut crops can, under certain conditions, result in contamination by toxic and carcinogenic mycotoxins, and it is important to ensure that primary produce contaminated with these toxins does not enter the food chain. Quantitative NMR is particularly useful for materials of high toxicity and/or limited availability because a fit-for-purpose assignment can be obtained using a few milligrams of material.

NMR measurements are complex, and the results are influenced by factors including sample preparation, signal generation, and the design of the NMR experiment used for quantification. A detailed study of the measurement process led to the Ishikawa (fish tail) diagram shown below; each factor could then be assessed in turn leading to a robust estimate of the overall measurement uncertainty of the assigned value for mass fraction content of the material.

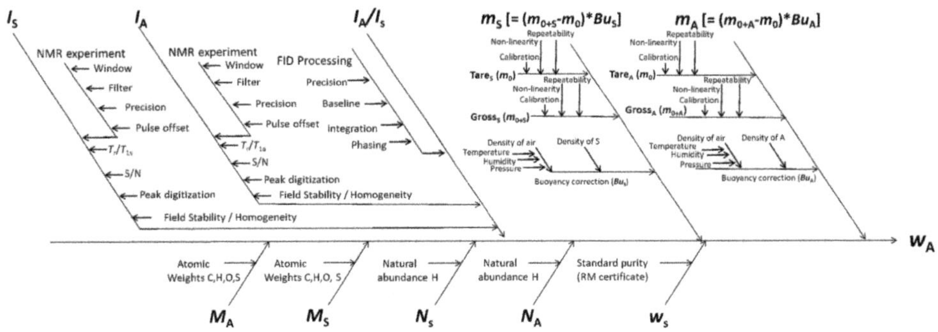

Further details and an explanation of each of the factors are available in a BIPM report [15].

Drs Steven Westwood, Gustavo Martos, and Robert Wielgosz, BIPM, Paris

the uncertainty can be estimated. Type A uncertainties can be determined by statistical analysis of repeated measurements. Sometimes this information can be found in historical data or the physics of the measurement process (for example, in radioactive decay, the statistical distribution of the decay rate is known), but it normally involves carrying out experimental studies such as those for validation.

The question can come up about what exactly is meant by repeating measurements—for example, if your measurement involves taking a subsample from a sample of a pharmaceutical, do you need to take repeated subsamples or can you remeasure the same subsample? The answer is that normally repeating measurements means following the full procedure for each measurement; in this case you would take another subsample. This might not be possible in practice so you may need to include an additional 'Type B' uncertainty to account for variations in subsampling—the aim is to include all the important components of uncertainty somewhere in the uncertainty budget.

Another issue for Type A uncertainties is that the repeated measurements are normally carried out over a short time period, so the effects of variations in laboratory conditions are not included. An additional Type B component may be needed for these effects.

Type B uncertainties are components that are estimated by techniques other than statistical methods. Many sources of information can be used to determine Type B uncertainties, including:

- Information from calibration certificates for instruments or reference materials
- Publications of any reference data used
- Scientific publications (articles, conference proceedings, proficiency test exercise reports)
- Documentary standards and good practice guides
- Instrument manuals
- Analogy with other measurements
- Mathematical models

The difference between Type A and Type B uncertainties is the method used to estimate them, a mathematical rather than a scientific distinction. As the GUM points out, Type A uncertainties are not necessarily more reliable than Type B uncertainties: there is an uncertainty in the uncertainty of a Type A component; if the standard deviation is calculated from ten replicates, the uncertainty in the standard deviation is 24%.

Given this list of Type A and Type B uncertainties, the next challenge is to express the uncertainty in each input parameter (a_1 etc) as a standard deviation—the 'standard uncertainty' (u_1 etc). This is easy to calculate for Type A uncertainties for which the standard deviation can be calculated from repeated measurements. Estimating the standard uncertainty for Type B uncertainties depends on the component: for example, the standard uncertainty is stated on calibration certificates but may need some estimation and judgement for other components. It depends on the application; estimating the Type B uncertainties can be a major part of the work.

The estimate of the Type B uncertainty should be refined to take account of the probability distribution (the probability density function) of the input parameter. For example, you may know that an input parameter must lie within a particular range with equal probability—the standard uncertainty should be calculated taking this into account. The most common probability distributions are shown in figure 2.20.

Judgement will be needed to select the probability distribution—it is important in some cases (for example, realizing primary standards at a national metrology institute) and less important in other cases (the different components are combined at the end of the process, so the shape of the probability distribution of individual components may not be significant in the final result).

Distribution	Shape of probability distribution (probability density function)	Characteristics	Standard uncertainty	Example
Rectangular	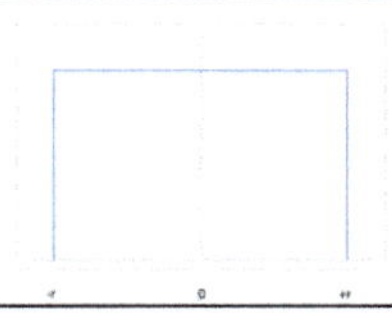	The input parameter can take any value between a lower (-r) and upper limit (+r) with equal probability	$\frac{r}{\sqrt{3}}$	The resolution of a reading taken from a digital display
U-shaped	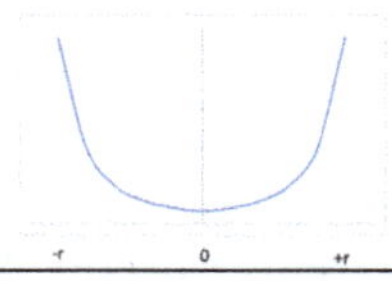	The input parameter is most likely to take a value at the extremes.	$\frac{r}{\sqrt{2}}$	Laboratory temperature (triggered by a thermostat at either end of the range)
Triangular	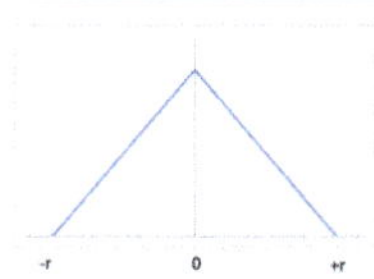	The input parameter is most likely to have a value at the centre of the distribution but is more likely have a value away from the mean than a normal distribution.	$\frac{r}{\sqrt{6}}$	Sometimes called the 'lack of knowledge' distribution, it is used when there is limited information about the distribution – it is more cautious than assuming a normal distribution.
Normal distribution	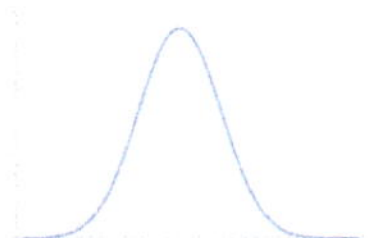	The familiar bell-shaped distribution.	Standard deviation	Usually assumed for components such as imported uncertainties from calibration certificates.

Figure 2.20. Common probability density functions.

The next step is to calculate how sensitive the result of the measurement is to changes in the input parameter. From the measurement equation, a small change in one of the input parameters (say, a_1) will change the result by:

$$\Delta\text{Result} = \frac{\delta F}{\delta a_1}\Delta a_1 = c_1 \times \Delta a_1$$

The sensitivity coefficient c_i is defined to be the partial differential of the measurement equation with respect to the input parameter a_i. For example, suppose the temperature in the laboratory is controlled to 20 ± 1 °C—this does not mean that a 1 °C change in temperature changes the result by 5%—the impact on the result may be much less, and the sensitivity coefficient takes this into account. The sensitivity coefficient may have to be determined experimentally, in this example by carrying out replicate measurements at different temperatures and calculating the sensitivity coefficient from:

$$c_n = \frac{\Delta\text{Result}}{\Delta a_n}$$

You now have three pieces of information about each component of uncertainty, to be captured in the 'uncertainty budget':

- The type of uncertainty (Type A or Type B)
- The standard uncertainty (u_i), calculated from statistical analysis or other methods

- The sensitivity coefficient (c_i), calculated by taking the partial differential of the measurement equation or by experiment

The combined standard uncertainty u_c can then be calculated using the law of propagation of uncertainties, by multiplying each standard uncertainty by the sensitivity coefficient and adding them in quadrature:

$$u_c = \sqrt{\sum_i c_i^2 u_i^2}$$

This formula does not need to consider the shape (normal, rectangular, U-shaped, etc) of the probability density functions for the different components, as the standard uncertainty already takes this into account. The beauty of the approach used in the GUM is that all components of uncertainty can be treated in the same way.

One issue with developing an uncertainty budget is deciding how much effort to put in. Quality assurance standards require you to identify and quantify the main components, but the definition of 'main' depends on the application. There are always many possible influences on a measurement; the effort expended will depend on the possible impact. If you are calibrating an instrument or preparing a reference material, you may not know the potential impact, so looking into the uncertainty budget in detail is worthwhile—the same goes for work that affects health or has high cost implications. However, if you know that the uncertainty budget is overwhelmed by other factors (for example, the measurement of environmental samples which by their nature are heterogenous), a simplified uncertainty budget will probably suffice. I say 'probably' because it does depend on the application and there can be traps for the unwary.

If correction factors need to be applied to the output, it is not good practice simply to add an additional component to the uncertainty budget; instead, the correction should be applied and a component included for the uncertainty in the correction.

The uncertainty is presented as an interval and the level of confidence associated with the interval, by multiplying the combined standard uncertainty by a coverage factor k. Assuming a normal distribution, a coverage factor of $k = 2$ corresponds to a level of confidence of about 95%; that is, you are 95% certain that the 'true' value lies within $2 \times u_c$ of the result.

A full description of the result of a measurement is therefore:

- The result in the form $y \pm U$, where $U = k \times u_c$
- The units of y and U
- The coverage factor k
- The approximate level of confidence associated with the interval $y \pm U$

The number of digits in the result should not give a false impression of the accuracy of the measurement.

In practice, it is possible to simplify the estimation of the measurement uncertainty. There are also instances when the implicit assumptions are not valid. These two cases are discussed below.

Case 1: Simple calculation of uncertainty budgets

Combining uncertainties in quadrature means that the total uncertainty is usually dominated by one or two components. Details such as the precise probability distribution of the smaller components therefore have negligible impact on the combined uncertainty, so an informed guess of the probability distribution will normally suffice.

Some smaller components can be omitted from the calculation altogether. Information from the scientific literature or other evidence may enable you to say that the uncertainty from that component is less than xx% and is negligible.

It is also possible to avoid using differential calculus if the measurement equation is just multiplying or dividing factors, such as:

$$Y = A \times B/C$$

Then, from the law of propagation of uncertainties, the combined relative standard uncertainty of Y (as a percentage) is given by adding the relative standard uncertainties of A, B, and C in quadrature:

$$u_Y(\%) = 100 \times \frac{u_Y}{Y} = 100\sqrt{\left(\frac{u_A}{A}\right)^2 + \left(\frac{u_B}{B}\right)^2 + \left(\frac{u_C}{C}\right)^2}$$

Many measurement equations can be expressed in this form.

Another common form is:

$$Y = A - B$$

which is typical if one is subtracting a background signal. Working through the mathematics, the standard uncertainty of Y is given by:

$$u_Y = \sqrt{u_A^2 + u_B^2}$$

These approaches may simplify the mathematics needed, but, from a quality management system point of view, they still have to be justified. The uncertainty may be dominated by one component that is out of your control (a common example is wide variations between environmental samples) but you have to demonstrate that the measurement process itself is under control and understood, so estimating the measurement uncertainty is still needed.

Case 2: Complex uncertainty budgets

The first complication is that the formula for the propagation of uncertainties above assumes that the components of uncertainty are independent of each other. This is sometimes not the case; as an example of how this can go wrong, suppose you want to find out the amount of an analyte in a sample. You decide to use two different techniques to measure the sample; you get two results, A and B, and take the mean.

The measurement equation has the form:

$$Y = \frac{A + B}{2}$$

From the law of propagation of uncertainties, the standard uncertainty of Y is:

$$u_Y = \frac{1}{2} \times \sqrt{u_A^2 + u_B^2}$$

Suppose that the same reference material is used to calibrate the two methods, so both results are multiples of the reference value R:

$$A = a \times R \text{ and } B = b \times R$$

If the uncertainty in the reference value R is certificated to be 5% and the uncertainties in a and b are negligible, the combined standard uncertainty in Y is 3.5% according to the law of propagation of uncertainties. This implies that your measurements could be used to recertify the reference material with a lower uncertainty, and the process repeated until the uncertainty in the reference material is zero! The combined standard uncertainty must be at least 5%—why is the calculation wrong?

The answer is that the uncertainties in A and B are correlated. There are two ways to deal with this problem. The first way is to rewrite the measurement equation so that all the components are independent:

$$Y = \frac{1}{2} \times (a \times R + b \times R)$$

There is now an additional component in the measurement equation for the reference material, and the usual formula is correct—the combined standard uncertainty is 5%. The same approach can be used for more complex measurement equations, working through the calculations in stages ensuring that components of uncertainty are combined only if they are independent.

The second way is to use the full equation for the propagation of uncertainties from the GUM:

$$u_c^2 = \sum_i c_i^2 u_i^2 + 2\sum_i \sum_{j,j \neq i} c_i c_j u(a_i, a_j)$$

The second term in this equation is associated with the correlation between the input parameters a_i and a_j. For Type A uncertainties, the covariance (for parameters a_1 and a_2, say) can be calculated by taking n independent pairs of simultaneous observations:

$$u(a_1,a_2) = \frac{1}{n-1}\sum_{j=1}^{n}(a_{1j} - m_1)(a_{2j} - m_2)$$

where

$$m_1 = \frac{1}{n}\sum_{j=1}^{n} a_{1j} \quad \text{and} \quad m_2 = \frac{1}{n}\sum_{j=1}^{n} a_{2j}$$

For Type B uncertainties, the equation above can be written in terms of a 'correlation coefficient', r:

$$u_c^2 = \sum_i c_i^2 u_i^2 + 2\sum_i \sum_{j,j\neq i} c_i c_j r(a_i, a_j) u_i u_j$$

The correlation coefficient is positive if a positive change in one parameter means the second parameter will also change positively, and negative if the second parameter changes negatively. The correlation coefficient is zero if the parameters are uncorrelated. In the simple example above, the correlation coefficient is +1 and the combined standard uncertainty is 5%, as expected.

The second complication is that in estimating the measurement uncertainty, one is combining information on the probability density function of each input parameter to arrive at a combined probability density function for the output parameter. The law of propagation of uncertainties is a special case of this: it is an approximation combining the standard uncertainty associated with each input parameter to determine a standard uncertainty associated with the output parameter. This is good enough for most applications, but there are examples where it does not work well—if the probability distributions of the input or output parameters are asymmetric, for example. One example with real-world consequences is the radioactivity content of contaminated material—samples can have a log-normal distribution for their radioactivity content; the distribution is approximately Gaussian in shape but with a long tail with more high-activity samples than might be expected. Ignoring the shape of the probability density function can result in a decision that the waste is more dangerous (and hence more expensive to dispose of) than is warranted.

There are two methods one can use to determine the probability density function of the output parameter. The first is to derive the exact solution in algebraic form using the measurement equation—this can result in a jungle of algebra from which few emerge unscathed. A more practical alternative is to use a Monte Carlo technique. This method is particularly useful if the measurement equation is very complex or if it is not easy to derive an analytical function (for example, if one input to the measurement equation is calculated by a software package).

In the Monte Carlo technique, a random number is generated for each input parameter and transformed to give a value that depends on the probability density function of the input parameter. The output parameter is then calculated from this set of input parameters. The process is repeated millions of times and the probability density function for the output parameter derived. The accuracy of the probability density function depends on the number of repetitions of the process. Strategies for using the Monte Carlo technique are described in detail in reference [17].

Reducing measurement uncertainty

There is one more benefit of studying measurement uncertainty in detail: the opportunity to reduce the uncertainty. This can help reduce the risk to your organisation (in particular, for conformity assessment—see the next section), give you more confidence in your results, make your work more valuable to your customers, and even uncover some new, publishable, scientific knowledge. You could even decide to change the method you are using.

As components of uncertainty are largely combined in quadrature, the place to start is the larger components. Type A components may be reduced by taking more replicate measurements, but there can be diminishing returns as long-term drifts in results become apparent. Type B components depend on the technique but may include more accurately calibrated instrumentation or reference materials, refining the mathematical model, or carrying out further studies to investigate components in more detail rather than overestimating the component.

2.6 Pass or fail decisions: conformity assessment

Another issue covered by the quality system is taking pass/fail decisions on results from measurements. For example, you may be asked to decide whether the content of an impurity in a pharmaceutical is greater or less than a limit set by the specification for the drug, a decision that has consequences for patient safety and financial implications for your company. Taking this decision is called 'conformity assessment' and may be required by your customer or a regulatory body.

The main terms used in conformity assessment are:

- 'Tolerance limit'—the specification that must be met, set by regulations, documentary standards, guidance, or the customer
- 'Decision rule'—the process, usually agreed with the customer, that will be followed to test if the results are within the tolerance limit
- 'Acceptance limit'—the result is a pass if it is within the acceptance limit, which may be different from the tolerance limit
- 'Statement of conformity'—a written statement of whether the result is a pass or fail

The scientific challenge of assessing conformity is that the uncertainty of measurement means that there is always of risk of making a mistake: it is possible to falsely accept or falsely reject a result [18, 19]. Decision rules must take measurement uncertainty into account, and this means that conformity assessment can be complex (the mathematics of conformity assessment is explained in detail in reference [20]).

The simplest option is called 'a simple acceptance decision rule'; the acceptance limit is the same as the tolerance limit with a precondition that the measurement uncertainty is less than a defined amount. The constraint on the uncertainty is called the 'measurement capability index' (*C*) [19] and has the form (for a two-sided test, where the specification sets an upper and lower limit):

$$C = \frac{T}{4u}$$

where T is the tolerance interval and u is the standard uncertainty in the measurement.

Similarly, the capability index for a one-sided test is:

$$C = \frac{T}{2u}$$

Values of C between 2 and 5 are typical, depending on the application.

In other words, if the maximum amount of impurity in a pharmaceutical is less than 10 mg, the result is assessed to pass if the content is less than 10 mg and the uncertainty in this measurement is better than 5%. The probability of making an error (false accept or false reject) will be high if many results are near the tolerance limit or measurement uncertainty is high. Conversely, the probability of making an error will be low if results are spread out over a wide range or if the measurement uncertainty is low.

A more cautious option is to define a 'guard band', an approach that is sometimes enforced by regulators. The acceptance limit is defined to be smaller than the tolerance limit by an amount that depends on the typical measurement uncertainty —this guard band is usually two times the uncertainty, but wider guard bands are used if the consequences of false acceptance are severe. Figure 2.21 shows the difference between a simple acceptance decision rule and a guard band.

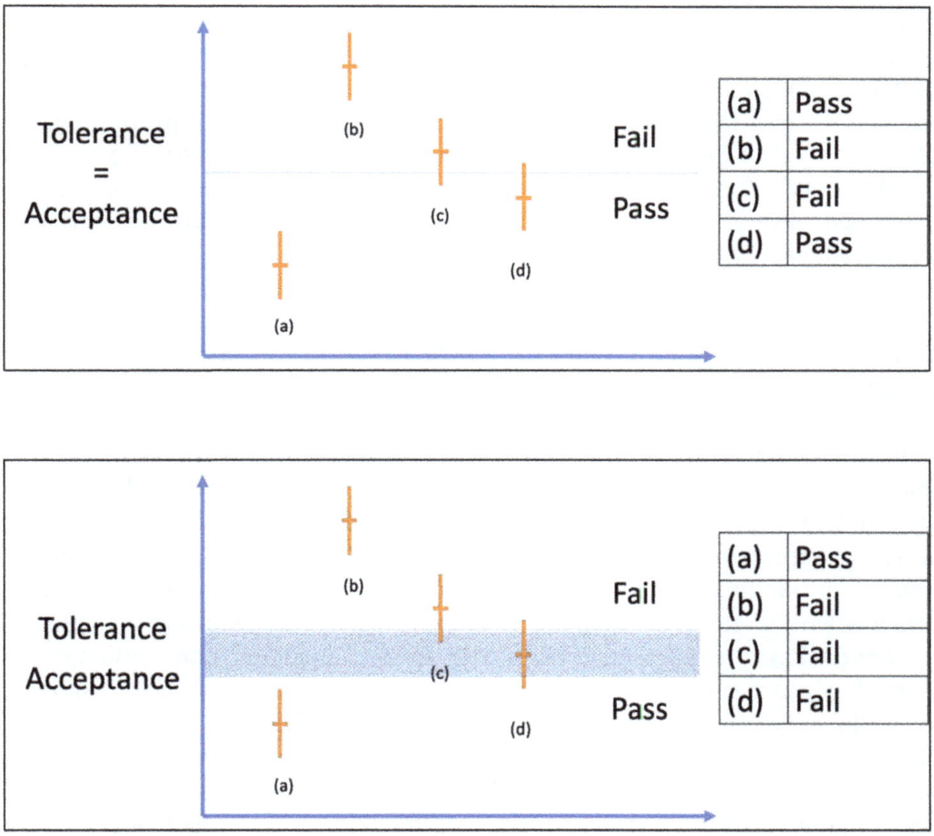

Figure 2.21. Decision rules for conformity assessment. Part (a) shows the simple decision rule in which the acceptance limit is equal to the tolerance limit (a one-sided limit): results (a) and (d) pass, although the uncertainty from result (d) shows the true result could be a fail. Part (b) shows the effect of introducing a guard band (shaded); result (d) is now assessed as a fail.

A result may also be assessed as a 'conditional pass' if it is within the guard band or a 'conditional fail' if it is outside the acceptance limit but the measurement uncertainty (at a given level of confidence) overlaps the acceptance limit.

2.7 Software validation

Many scientists have experience of developing software for instrument control, data collection, and analysis, and it can be an enjoyable and rewarding part of the job. Demonstrating that the software meets a specification and is reliable can be challenging, overlapping with the discipline of software engineering. It can also be difficult to retain control over the software in the long term and to understand exactly what the software does; there are all too many examples of errors, including the serious consequences of delivering the wrong radiation doses for cancer therapy[3].

Most, if not all, scientific studies and applications rely on software for data acquisition, instrument control and automation, processing and archiving data, and analysis (modelling, simulation, and data visualization). Quality systems require software to be validated, documented, and protected, but little detail is given in standards such as ISO 9001 or ISO 17025. Some software is critical to safety and there can be strict regulations to follow. One comprehensive example of guidance is *GAMP 5: A Risk-Based Approach to Compliant GxP Computerized Systems* [21] (GAMP stands for 'Good Automated Manufacturing Practice'). This guide is maintained by a technical committee of the International Society for Pharmaceutical Engineering and is intended to help pharmaceutical companies comply with the US regulations, but it is used worldwide. The issue of software validation is so important that the US national metrology institute (amongst others) has a software quality group and the BIPM has formed a task group to develop straightforward guidelines for scientists.

The regulations and guidance on software validation are constantly evolving, so the aim of this section is just to raise awareness of the issues to be considered and provide some basic guidance. Perhaps it is no surprise that some organizations therefore have specialist teams responsible for developing and validating software, and you are not allowed to implement your own. The approaches below are only general guidance of issues to consider; the requirements in your organization may be very different.

In a scientific environment, you may be expected to purchase a commercial software package (perhaps bundled in with an instrument), to commission custom-designed software from a supplier, to develop a spreadsheet to calculate results from input data, or to write your own code for instrument control and data analysis. In all cases, you will need a specification, evidence that the software meets the specification, instructions on using and maintaining the software, and a way to ensure the latest version is always the version in use. As time goes on, you will also need information on the hardware, compilers, and software libraries used by the code, to make it easier to update the code.

[3] For example, https://www.iaea.org/resources/rpop/health-professionals/radiotherapy/accident-prevention/errors-in-treatment-planning.

The next sections summarise some general guidance on software validation for different cases.

2.7.1 General guidance: validation of spreadsheets

Spreadsheets are probably used in every single laboratory. They are quick and easy to develop, are versatile, and have built-in tools to plot the results. They are often developed and used by a single person who knows about the spreadsheet's foibles and interprets the results accordingly. The problem is that the spreadsheet can become embedded in the organization's procedures, the original purpose and the constraints on its use forgotten. Spreadsheets combine data input, analysis, output, and presentation, so it can be difficult to follow their logic.

From a quality system perspective, creating a spreadsheet, however simple, is still software development and needs to be treated as such and subjected to some validation including a specification and tests. The depth of validation depends on the consequences of possible errors and an analysis of the risks.

The National Physical Laboratory's good practice guide for the development and testing of spreadsheet applications [22] points out many of the hidden problems in using spreadsheets and sets out recommendations for their use, including:

- Clear identification: details of the spreadsheet's name, purpose, authors, date of creation, version number, list of changes since creation
- Protection: so that only authorized members of staff can make changes, and data can only be entered in certain cells
- Documentation: instructions and references to equations and algorithms, embedded in the spreadsheet or in a separate document
- Set out clearly: using an agreed convention for layout
- Help for future developments: comments in the spreadsheet or within the text of any macros used
- Checks on input data: using cell formatting to prevent some data entry errors
- Checking calculations: using tracing tools to confirm that all the cells work as expected
- A 'delete data' key: a way to clear previous data, to avoid the risk of combining data sets in error

Validation of the spreadsheet should include checking the accuracy of the results using test data, testing any logic paths to check that they work, testing each function, and checking what happens if data outside the normal range is entered into the spreadsheet. Macros are a special case and should be treated as 'in-house software' and validated accordingly (see the next section). If calculations are complex and it is not feasible to check the results by hand, mathematics software packages can be used. A second suitably qualified and experienced person should also check the spreadsheet.

Maintaining a spreadsheet can be a problem, as it may not be compatible with new versions of the spreadsheet software. It can be safer and more robust to use a spreadsheet as a prototyping tool and code the software using a programming language for the final version.

In summary, spreadsheets are incredibly useful; the most important rule perhaps is to keep them simple so that they can be tested and maintained. The few simple precautions listed above should help ensure that a spreadsheet can be included in the quality system and will pass audits.

2.7.2 General guidance: in-house software

It used to be common practice in science to write one's own software using programming languages such as BASIC or FORTRAN; there was little choice, as commercial software was not available. Development tended to be ad hoc with individual scientists working on their own programmes. Since then, software development has been transformed, with new object-orientated programming languages that are more flexible but more complex, and more formal approaches have been adopted to ensure compliance with the quality system.

A software development life cycle is a step-by-step approach for software development. It is described in detail in ISO/IEC/IEEE 12207 [23], but that publication is intended for large-scale software development. There are several different models for the life cycle, including the V-model, but also models that support iterative development of the software. The main steps in a software development life cycle are shown in figure 2.22.

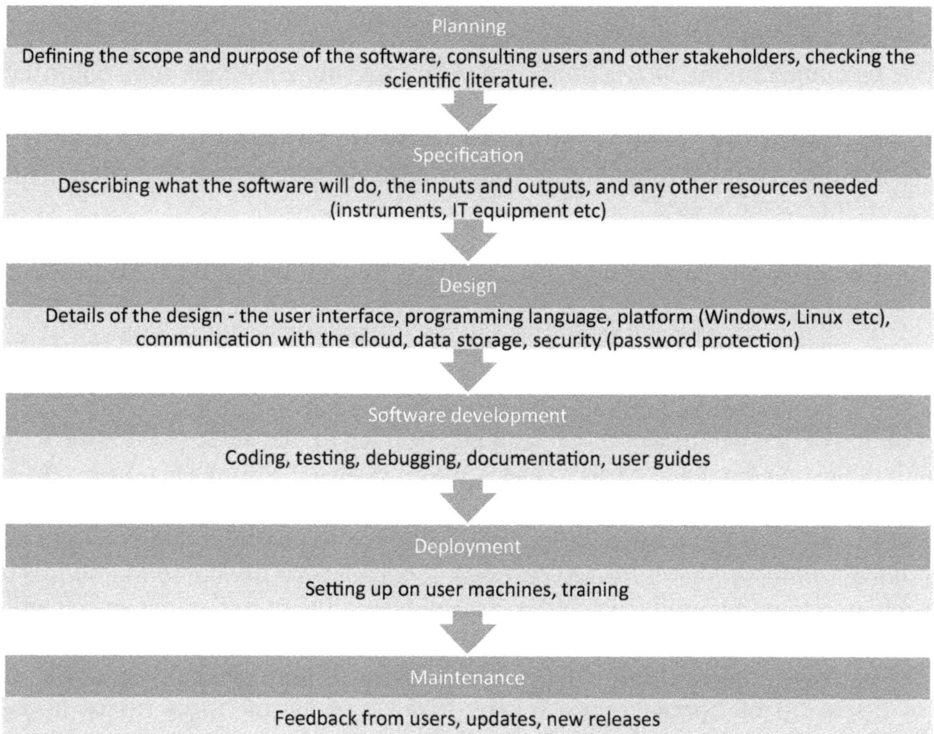

Figure 2.22. Software development life cycle.

From a quality assurance perspective, there are some issues to consider if you do decide to develop in-house software. Most importantly, ensure that it is possible for someone else to follow the logic of the code, so that it can be checked: keeping the coding style simple and modular, using names of variables that are easy to understand, and adding comments throughout are obvious steps (one in-house software package I inherited had two comments embedded—'start' and 'end'—debugging was a nightmare).

Version control is particularly important for in-house software, keeping track of changes and ensuring that you are always using the latest version. Keeping track manually is time consuming—software engineers use development platforms (github.com is one example at the time of writing) to control the code. These platforms are free of charge, and time invested to learn how to use them can pay dividends in the long term.

There are other tools available to support in-house software development, systems such as Doxygen that will extract all the embedded text in the programme to create an up-to-date user manual avoiding the need to write a separate document. Debugging tools are available to track down any problems.

Testing of in-house software usually has three levels. First, unit testing—checking that each module of the software converts the inputs to the required outputs. Second, integration testing in which modules are combined into larger groups and the results compared to the Functional Specification. Third, system testing in which the full software package is tested against the User Requirements Specification.

Similar issues apply to using tools such as LabVIEW or adaptions to the public-domain software packages found in the scientific world (Monte Carlo simulation packages for radiation dosimetry being one example).

2.7.3 General guidance: commercial software packages

Most instrumentation is supplied with software to control the equipment and to analyse the results. The software is essentially a 'black box' and it is not possible to validate each module. There are however some steps that can be taken to reduce the risk of errors and demonstrate to auditors that the software is under control. It can be tempting to assume that the manufacturer's software will have no problems, but experience has shown this is unwise; manufacturers can lose track of source code or lose the staff who wrote the code, leaving the user with nowhere to turn for support.

When purchasing a new instrument, the first step is to look for a supplier that complies with a software development standard such as ISO/IEC 25000 [24] or is accredited to ISO 9001; this will give some confidence that the supplier has its own systems for developing, maintaining, and documenting the software. It is also possible to ask the supplier for evidence of validation of the software, to have copies of the results from tests, or to audit the supplier in person. Some suppliers may object on the grounds of commercial confidentiality, but the issue of validation can be so important in industry for commercial or safety reasons that it can be made a condition of purchase and arrangements negotiated that will satisfy both parties.

The next step is to treat the software as an integral part of the instrument and follow the general validation procedures described above, testing the software by measuring known samples, challenging the inputs and outputs at the limits of the range, checking reproducibility, etc. The software will normally be supplied with a user manual, but an operating procedure will still be needed to describe how the software is used in practice.

A further issue with commercial software packages is that changes to the software are not under the control of the purchaser. Updates are issued regularly, and a procedure will be needed to decide whether to adopt the new version and what validation should be carried out.

2.7.4 General guidance: bespoke software

If no commercial software is available and writing the software oneself would be onerous, the software has to be commissioned from a specialist service in another part of the organization or another company.

To ensure that the custom software is under control, an important step is to devise a very detailed specification, far more detailed than if you were writing the software yourself. Mock-ups of screens, tight descriptions of the inputs and outputs, details of calculations, and flow charts all help. The more detail in the specification, the better the chance that you will get what you want—the supplier may not be familiar with the jargon in your field and there is a risk of misunderstanding.

The detailed specification will make writing a test plan easier to demonstrate that the software works as expected. A record will be needed of any bugs found and the corrective actions taken.

2.8 Demonstrating that the validation studies have worked—proficiency test exercises

All the work to validate and estimate uncertainties has one major weakness—it is carried out in isolation within your organization and depends on the skills, knowledge, and experience of a few individuals. There are risks of errors in implementing the procedures, possible bias in results, and you may be adopting procedures that other organizations have shown to have problems; your customers, regulators, and accreditation bodies may want more evidence of the quality of your work. An independent method to check results against reference values or evaluate performance against other laboratories is needed.

Many organizations therefore participate in 'proficiency test exercises' (PTEs). An independent pilot laboratory provides an artefact or an instrument; your laboratory carries out measurements on the artefact or the calibration factor for the instrument and returns the result to the pilot. The pilot collates the results from the participating laboratories and publishes a report. The report shows how close your result is to a consensus value and you can also assess if your estimate of measurement uncertainties is reasonable. Participating in PTEs is normally expected for laboratories accredited to a quality standard such as ISO 17025—just part of the routine business of the laboratory. Participation may be compulsory if the laboratory is part of an international project and the coordinating body wishes to ensure consistent measurements.

Accreditation bodies have published guidance to help organizations decide how often to participate in PTEs. The organization has to take into account the probability and impact of errors, considering issues such as how results are used (forensic science for example would imply a high level of participation in PTEs), number of measurements, the complexity of methods, staff experience, and staff turnover. Participation can be expensive and time consuming, so other methods to check results such as regular use of reference materials and in-house comparisons with results from other techniques can help reduce the frequency needed. In some fields, the frequency is constrained by availability of the exercises.

For a scientist, taking part in a PTE is a challenging part of the work, as the results will be compared to others (including competitors in some cases) and no one wants to be an outlier or to lose membership of an international network. The compensation is that there are many benefits: a good result (i.e. your result agrees with the 'correct' value within the measurement uncertainties) from the PTE demonstrates the quality of work to your management, to quality auditors, to regulators, and to customers. It also gives you and your colleagues a sense of pride in your work and contributes to the recognition of your laboratory in the community.

Conversely, a 'bad' result can alert you early to any problems, help you identify improvements, and find possible issues with equipment, calibrations, methods, and operator training. These are issues that are better addressed in the context of a PTE rather than following customer complaints.

There are other advantages just from participating. You can compare the results from different methods or different staff involved within your laboratory, even if you only submit one of the results to the pilot laboratory for the exercise. Some pilots analyse the results in detail and include a critical assessment of the advantages and disadvantages of the different techniques used, which may change your working practices; the reports often include more detail on estimating uncertainties than can be found in the scientific literature. Many PTEs include a workshop as an opportunity to discuss the results with your peers and seek solutions to common problems. Finally, irrespective of the result, participation shows a commitment to continuous improvement and encourages learning as the outcome will be mulled over by others in your organization.

Finding a PTE depends on the sector in which you work. Accreditation bodies such as the United Kingdom Accreditation Service maintain a list of providers—PTE providers are accredited to the ISO 17043 quality standard [25]. Many national and international institutes also offer PTEs. If you can find nothing in your field, an informal comparison exercise can be easy to organize with another laboratory.

The rules for participation depend on the exercise but will address issues such as:

- The timetable for participation
- Deadlines for returning results
- The format for results and any additional information needed
- Confidentiality—no discussion of results with other participants

- Anonymity—your laboratory's results may be visible or identified by a code only known to you and the pilot
- Whether you can withdraw or correct results after submission (for example, corrections for typographical errors noticed after submission)

The pilot laboratory will also describe the artefacts or instruments in detail and the approach to data analysis and scoring. Your results will be compared to the 'true' value, but how that value is calculated depends on the exercise. In some cases, the pilot laboratory will assign a value based on its own measurements—for example, if the artefact is prepared from a traceable reference material, the pilot will have access to additional information that enables it to assign a value with confidence.

In other cases, the true value is determined by combining the results from all the participants to obtain a consensus—this is more difficult than it seems. Taking the arithmetic mean of results ignores the issue that some results have lower uncertainties than others. A weighted mean considers the uncertainties, but these uncertainties may not have been calculated correctly and the results may be inconsistent; some results may be wrong due to typographical errors and should not be included in the mean. Many different approaches have been proposed to calculate the 'true' value, including artificially increasing uncertainties so that results are consistent, using statistical tests to eliminate outliers, and setting the true value based on the largest consistent subset of the results (it is the same problem as analysing the results from international comparisons discussed in chapter 1). A value calculated by consensus may not be the true value—it could be that one technique is better than others in reality—but this is impossible to assess from the PTE and the consensus value is sometimes the best one can do. Whichever technique is chosen, the pilot laboratory will describe it before the exercise starts.

The pilot will also provide an assessment of how your result compares to the true value. Typically, the pilot will calculate:

- The deviation of your result (L) from the consensus value (N):

$$D = \frac{L - N}{N}$$

- The standard uncertainty u_D of the deviation, by combining the standard uncertainty your laboratory has provided (u_L) and the standard uncertainty of the consensus value (u_N):

$$u_D = \frac{L}{N}\sqrt{\left(\frac{u_L}{L}\right)^2 + \left(\frac{u_N}{N}\right)^2}$$

- Other quantities may be calculated to help assess how well your laboratory has done, such as:

  The zeta score:

$$\zeta = \frac{L - N}{\sqrt{u_L^2 + u_N^2}}$$

The Z score:

$$Z = \frac{D}{\sigma_P}$$

The pilot laboratory then sets a pass/fail value for the zeta and Z scores. The zeta score determines whether your result is significantly different from the consensus value based on your estimate of the measurement uncertainty. The Z score is similar but uses a standard uncertainty for the exercise (σ_P) set by judgement by the pilot laboratory—it depends on the application of the PTE and likely measurement uncertainties. The Z score ensures that laboratories cannot pass a PTE by artificially inflating the measurement uncertainty.

As a further check to confirm that uncertainties are comparable, the pilot may also publish the relative uncertainties estimated by participants and use a statistical outlier test to identify the results that pass.

The pilot laboratory may also classify the results as pass (both zeta score and Z score passed and relative uncertainty not rejected as an outlier), questionable (either the zeta score or Z score failed), or fail (both zeta and Z score failed). The report also normally includes a plot that shows how closely the results from all participants lie to the true value.

The real value of participation in a PTE comes from looking carefully at your result and asking if it is fit for purpose and whether there are any improvements you could introduce or refinements to the uncertainty budget. Participating in PTEs over a long period will show you how the quality of your work gradually improves. There is long-term benefit for the whole community in participating in PTEs. Participants review the results from the PTE and refine their procedures, correcting errors that may have happened in the past, gradually leading to more consistent measurements.

2.9 Maintaining the quality of measurements

Validation happens at a moment in time; it may need to be repeated if a new method is adopted or if an instrument breaks down and is repaired or replaced, but normally it is a one-off exercise. This is not enough for a quality management system; it is also important to demonstrate that nothing has changed since the validation was completed.

There are two aspects to maintaining the quality of measurements: keeping the equipment under control (see Section 2.2) and demonstrating consistent results.

2.9.1 Demonstrating consistent results—control charts

The second aspect is obtaining evidence that the measurement system, whatever it may be, is delivering consistent results. A side benefit of assembling the evidence for consistent results is that it shows that you understand the system and have thought about potential problems.

One tool you can use is a control chart. This is a chart of the main parameters that could affect the results over time. The idea is to apply the well-known properties of a normal distribution to determine whether deviations in a parameter are within acceptable limits; essentially a control chart is a normal distribution turned on its side with time as the x-axis. Plotting the results enables you to spot any long-term trends or discrepancies quickly, and statistical analysis helps you identify more subtle problems you may miss.

The statistical basis of control charts was developed in 1924 by an American physicist, Walter Shewhart, working at Bell Laboratories [26]. The background to Shewhart's work is an interesting story in itself; he joined Bell Laboratories at the time when there was a national drive to enhance productivity and innovation and when statistical methodologies were being developed globally. The control charts enabled significant improvements in production but had to overcome resistance, as people did not want to spend time collecting and analysing data and investigating the causes of discrepancies (this issue has not gone away!). Despite this, the benefits were clear; control charts gained acceptance and their use was spread through publication of books, articles, and higher education courses.

Control charts are nowadays described in detail in a series of ISO standards (the ISO 7870 series [27]). Applying control charts can be complex, but a simple approach can be suitable for most cases. The first step is to choose the parameters to track, for example:

- The response of an instrument to the same reference material or a reference signal
- 'Noise' on a system—for example, by recording the full-width half-maximum of peaks on a spectrometer
- The room temperature and humidity

The measurements should be easy to carry out and limited to the key parameters or instruments. The whole measurement process does not have to be carried out; one can track the output from particular instruments using techniques to mimic the inputs (a signal generator, for example).

For each parameter, the usual approach is to start by collecting 11 data points, allowing enough time between each measurement. Calculate the mean and standard deviation, and mark on the chart a 'warning limit' of mean ± 2 standard deviations and an action limit (± 3 standard deviations). This gives you a starting point for the control chart. Once you have accumulated 31 data points, recalculate the mean and standard deviation, and these are the levels to be used. A common mistake with control charts is to use a rolling mean—to continually update the mean and standard deviation as more data are added. This approach negates the purpose of a control chart as taking a rolling mean automatically compensates for any long-term drifts in results.

The straightforward way to use control charts is to rely on the warning and action limits. A data point outside the 'warning limit' will happen by chance once every 20 data points. The equipment is probably still OK and can continue to be used, but, if it happens again in the next 20 points, possible causes should be investigated. A data point outside the action limit will happen by chance 3 times every 1000 data points, so this indicates there could be a problem and possible causes should be investigated. In both cases, the results of the investigation should be recorded, even if it is just a note in the laboratory notebook. A typical control chart is shown in figure 2.23.

Control charts can also be used to look for trends and other issues:

- Nine successive data points all above or all below the mean (this can indicate that something in the system is shifting from one state to another)
- Six successive data points increasing or decreasing
- Fourteen successive points alternating up and down
- Eight successive points between one and two standard deviations of the mean
- Fifteen successive data points within one standard deviation of the mean

In all of these cases, possible causes should be investigated. They can give an early warning that something is deteriorating—for example, the insulation on a resistor in a high voltage system can start to break down; a warning that this is happening can enable you to arrange maintenance.

There are some problems with Shewhart charts in practice. First, the approach assumes that measurements are independent of each other, but this may not be true

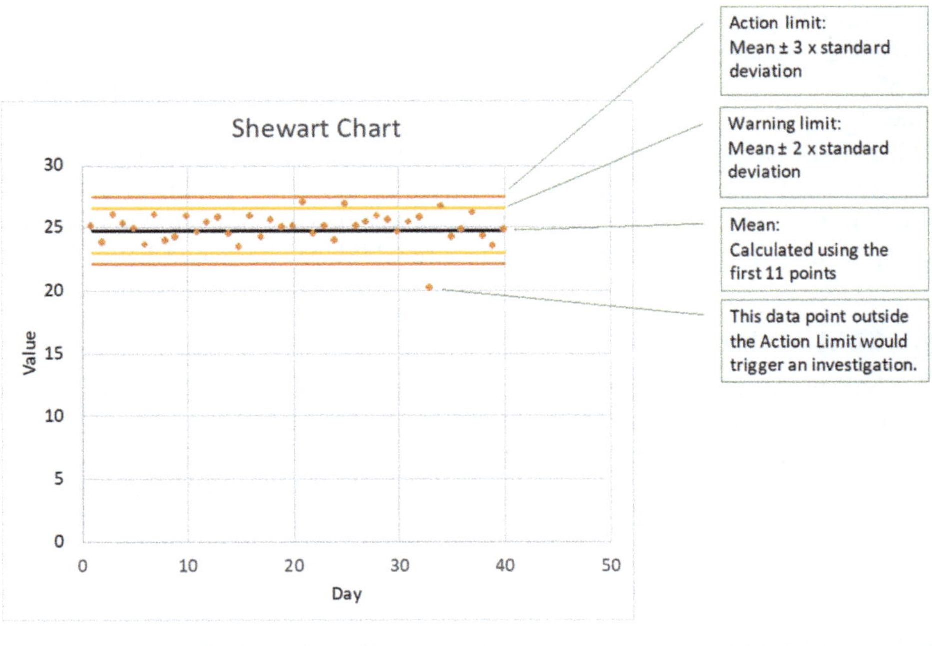

Figure 2.23. A typical control chart, showing the value of a parameter (the response of the instrument to a test artefact, for example) versus time.

if the time interval between measurements is less that the duration of a factor that influences the measurement. For example, the temperature of the laboratory is likely to be influenced by the temperature the previous day and the previous week. This can lead to frequent violations of warning and action limits, reducing confidence in the control charts. One way to avoid this issue is to take measurements less frequently to reduce the correlation between measurements but this has the risk that real problems will be missed. Another pragmatic approach is to consider the influence that the parameter has on the result of the measurement and set warning and action levels based on its impact. More sophisticated techniques exist such as the autoregressive integrated moving average (ARIMA) control chart[4] which drops the assumption that data points are independent; a model is used in which a data point depends on previous data points.

If there are frequent warning and action limit violations, it's worth investigating why and setting limits based on the impact on the measurement. There is the opposite problem that the limits are too wide, in which case they should be reset using a different set of 31 data points. Control charts can also be overinterpreted—not every trend is a problem, or underinterpreted—they don't show you all the problems.

There are two more questions to answer about Shewhart charts—when to start using them and how often to collect the data points. The answer to the first question is 'as soon as possible and during the validation process' so that it can be shown that nothing has changed since validation. The answer to the second question is that 'it depends'. The results may show that the process is under control over a long period and increasing the intervals between checks would save time, but it depends on the application. The consequences of omitting a check on an instrument measuring radioactive discharges from a power station are significant (risks to health and the risk of prosecution), so frequent checks are important even if there is a low risk of errors.

There is a vast amount in the scientific literature on the topic of control charts and many different approaches (there is an excellent summary on the website of the United States's national metrology institute[5])—the best approach to use depends on the application. Two common approaches are CUSUM charts and EWMA charts.

2.9.2 CUSUM charts

Cumulative sum (CUSUM) charts were developed because Shewhart charts are not very sensitive to detecting small shifts in the mean, as the fluctuations around the mean can disguise a gradual shift. Four parameters are needed for these charts:

- The expected mean (m) and the standard deviation (s), calculated using results from when you know the system was stable

[4] See, for example, https://cdn2.hubspot.net/hubfs/402067/PDFs/ARIMA_Charts.pdf.

[5] Control charts explained in the *NIST/SEMATECH e-Handbook of Statistical Methods* [10]: https://www.itl.nist.gov/div898/handbook/pmc/section3/pmc31.htm.

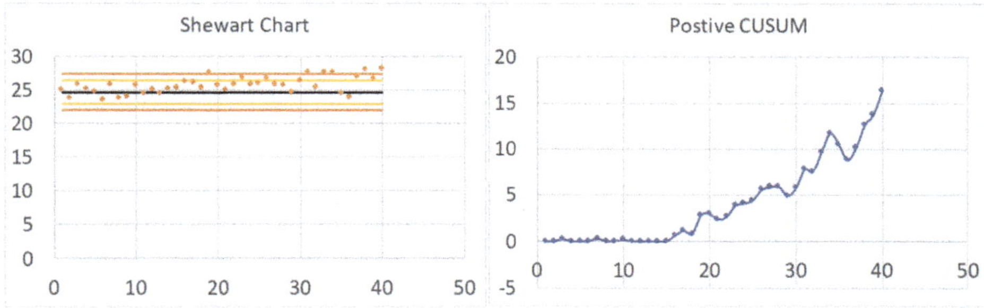

Figure 2.24. A comparison of a Shewhart chart and a CUSUM chart for the same data set (the y-axis is the result for a parameter, the x-axis is time). The data were generated using a normal distribution with a standard deviation of 0.89, but with the mean increasing after data point number 11. The value of k was taken to be the standard deviation. The trend in the mean is visible from data point 15 onwards in the CUSUM plot but is not obvious in the Shewhart chart until later.

- The shift in the mean that you want to detect (k). Values of $k = s$ or $k = 0.5$ s are commonly used.
- A decision threshold (H) that triggers an investigation of whether something is going wrong. A value of $H = 5$ s gives a reasonable balance between false alarms and missing changes in the mean.

For data point x_i, the positive CUSUM is calculated from:

$$C_i^+ = \text{Max}\{0,\ C_{i-1}^+ + x_i - (m + k)\}$$

The starting value (C_0) is set to zero. If data points are systematically more than the mean plus the parameter k ($m + k$), the CUSUM increases, triggering action when it reaches the value of H.

The negative CUSUM is calculated in a similar way:

$$C_i^- = \text{Min}\{0,\ C_{i-1}^- + x_i - (m - k)\}$$

Action should be triggered once the negative CUSUM reaches $-H$. An example is given in figure 2.24 showing how a CUSUM chart can detect trends in the mean earlier than a Shewhart chart.

2.9.3 EWMA charts

Exponentially weighted moving average (EWMA) charts also indicate when a mean is gradually shifting. The first step is to calculate the mean (m) and standard deviation (s) at a time when you know the system is stable.

The EWMA for data point i is then calculated as follows:

$$EWMA_i = \lambda x_i + (1 - \lambda)EWMA_{i-1}$$

The parameter λ is a weighting factor, the depth of memory of the calculation. Setting $\lambda = 1$ uses only the most recent data point; lower values of λ include information from other data points, smoothing out the variations to make it easier to see gradual drifts in the mean. A value of $\lambda = 0.2$ is commonly used.

Values of $EWMA_i$ are then plotted against time, the centre line being the mean m. Similar to a Shewhart chart, action limits (L) are set either side of the mean where:

$$L = k \times s\sqrt{\frac{\lambda}{2 - \lambda}}$$

The parameter k is usually set to 3.

2.9.4 Summary of control charts

Shewhart started with a simple aim—to provide simple graphical tools to control the complex processes that had emerged during the second industrial revolution. The approach gained traction in industry and in scientific research, and many different sophisticated techniques have been developed for applications ranging from environmental protection to marketing. The advent of ‘big data’ is driving further developments.

For most applications (assuring the quality of a scientific procedure) a simple Shewhart-type control chart is normally sufficient. Many instrument software packages include an option to record a control chart.

References

[1] 2017 *General Requirements for the Competence of Testing and Calibration Laboratories* ISO/IEC 17025:2017 International Organization for Standardization

[2] 2015 *Quality Management Systems—Requirements* ISO 9001:2015 International Organization for Standardization

[3] National Academies of Science, Engineering, and Medicine 2019 *Reproducibility and Replicability in Science* (Washington, DC: National Academies Press)

[4] Baker M 2016 1,500 scientists lift the lid on reproducibility *Nature* **533** 452–4

[5] 2016 *General Requirements for the Competence of Reference Material Producers* ISO 17034:2016 International Organization for Standardization

[6] ILACOIML 2022 *Guidelines for the Determination of Recalibration Intervals of Measuring Instruments* ILAC G24:2022, OIML D 10 International Laboratory Accreditation Cooperation (ILAC) and International Organization of Legal Metrology (OIML) (https://ilac.org/publications-and-resources/ilac-guidance-series/)

[7] 2014 *The Fitness for Purpose of Analytical Methods: A Laboratory Guide to Method Validation and Related Topics* ed B Magnusson and U Örnemark (Gembloux, Belgium: Eurachem) Eurachem Guide2nd edn (https://www.eurachem.org/index.php/publications/guides/mv)

[8] ICH 2005 *Validation of Analytical Procedures: Text and Methodology* Q2(R1) International Council for Harmonisation of Technical Requirements for Pharmaceuticals for Human Use (ICH) (https://www.ich.org/page/quality-guidelines)

[9] MARLAP Workgroup 2004 *Multi-Agency Radiological Laboratory Analytical Protocols Manual (MARLAP)* US Environmental Protection Agency, US Department of Defense, US Department of Energy, US Department of Homeland Security, US Nuclear Regulatory Commission, US Food and Drug Administration, US Geological Survey, and National

Institute of Standards and Technology NUREG-1576, EPA 402-B-04-001A, NTIS PB2004-105421 (https://www.epa.gov/radiation/marlap-manual-and-supporting-documents)

[10] 2003, updated 2012 and later *NIST/SEMATECH e-Handbook of Statistical Methods* ed C Croarkin and P Tobias *et al* (Gaithersburg, MD: National Institute of Standards and Technology) (http://itl.nist.gov/div898/handbook/)

[11] EA Laboratory Committee 2022 *Evaluation of the Uncertainty of Measurement in Calibration* EA-4/02 M:2022 European Co-operation for Accreditation (EA) (https://european-accreditation.org/publications/ea-4-02-m/)

[12] Pommé S 2015 Typical uncertainties in alpha-particle spectrometry *Metrologia* **52** S146

[13] BIPM Joint Committee for Guides in Metrology (JCGM) 1995, revised 2008 *Evaluation of Measurement Data—Guide to the Expression of Uncertainty in Measurement (GUM)* (Sèvres, France: International Bureau of Weights and Measures) JCGM 100 (https://www.bipm.org/en/committees/jc/jcgm/publications)

[14] Bell S 2001 *Good Practice Guide No. 11: A Beginner's Guide to Uncertainty of Measurement* (Tedddington, UK: National Physical Laboratory) (https://www.npl.co.uk/gpgs/beginners-guide-measurement-uncertainty-gpg11)

[15] Westwood S, Stoppacher N, Garrido B, Huang T, Saito T, Un I, Yamazaki T and Zhang W 2018 *Internal Standard Reference Data for qNMR: Maleic Acid* Rapport BIPM 2018/01 International Bureau of Weights and Measures (BIPM) (https://www.bipm.org/en/publications/rapports-bipm)

[16] BIPM Joint Committee for Guides in Metrology (JCGM) 2020 *Guide to the Expression of Uncertainty in Measurement—Part 6: Developing and Using Measurement Models* (Sèvres, France: International Bureau of Weights and Measures) JCGM GUM-6 (https://www.bipm.org/en/committees/jc/jcgm/publications)

[17] BIPM Joint Committee on Guides in Metrology (JGCM) 2008 *Supplement 1 to the Guide to the Expression of Uncertainty in Measurement—Propagation of Distributions Using a Monte Carlo Method* (Sèvres, France: International Bureau of Weights and Measures) JCGM 101 (https://www.bipm.org/en/committees/jc/jcgm/publications)

[18] ILAC 2019 *Guidelines on Decision Rules and Statements of Conformity* ILAC G8:09 International Laboratory Accreditation Cooperation (ILAC) (https://ilac.org/publications-and-resources/ilac-guidance-series/)

[19] UKAS 2022 *Decision Rules and Statements of Conformity* LAB 4 United Kingdom Accreditation Service (UKAS) (https://www.ukas.com/resources/publications/laboratory-accreditation/)

[20] BIPM Joint Committee on Guides in Metrology (JCGM) 2012 *The Role of Measurement Uncertainty in Conformity Assessment* (Sèvres, France: International Bureau of Weights and Measures) JCGM 106 (https://www.bipm.org/en/committees/jc/jcgm/publications)

[21] ISPE 2022 *GAMP 5: A Risk-Based Approach to Compliant GxP Computerized Systems* (North Bethesda, MD: International Society of Pharmaceutical Engineers) 2nd edn (https://ispe.org/publications/guidance-documents/gamp-5-guide-2nd-edition)

[22] Barker R M, Harris P M and Parkin G I 2006 *Software Support for Metrology Best Practice Guide No. 7: Development and Testing of Spreadsheet Applications* DEM-ES 013 National Physical Laboratory (https://eprintspublications.npl.co.uk/3878/)

[23] 2017 *Systems and Software Engineering—Software Life Cycle Processes* ISO/IEC/IEEE 12207:2017 International Organization for Standardization

[24] 2014 *Systems and Software Engineering—Systems and Software Quality Requirements and Evaluation (SQuaRE): Guide to SQuaRE* ISO/IEC 25000:2014 International Organization for Standardization

[25] 2023 *Conformity Assessment—General Requirements for Proficiency Testing* ISO/IEC 17043:2023 International Organization for Standardization

[26] Bradford P G and Miranti P J 2019 Information in an industrial culture: Walter A. Shewhart and the evolution of the control chart, 1917–1954 *Inf. Cult.* **54** 179–219

[27] 2014–2023 *Control Charts—Parts 1–9* ISO 7870 International Organization for Standardization

Quality Assurance for Scientists and Engineers
A practical guide
Steven Michael Judge

Chapter 3

Management aspects of quality

The aim of this section is to set out an overview of the general requirements of quality management systems such as ISO 17025 [1] and ISO 9001 [2]. From analytical, calibration, and test laboratories through to highly regulated industries such as the nuclear industry or pharmaceutical production, the requirements are broadly the same. The idea is to give you an indication of the culture of quality-focussed organizations and to describe the different ways of working you may come across.

As well as the technical aspects described in previous sections, a quality management system extends to how the organization is structured and how it is run. Selection and training of staff, purchasing and running equipment, independent audits, controlling documents—all of these and more besides are covered by the quality system.

The entry point into the quality system is usually through two documents, the 'Quality Policy' and the 'Quality Manual'. They may be called something else, they may be combined in one document, but all quality systems will cover the same material.

3.1 The Quality Policy

The Quality Policy is a very simple document, a brief statement that sets out the purpose of the organization and confirms its commitment to comply with a particular standard or set of regulations. It is usually signed by the CEO or the head of the organization to show their personal commitment. It advertises the quality of the organization to customers, regulators, auditors, staff, and other stakeholders; the Quality Policy is sometimes in the public domain and may even be published on the website.

doi:10.1088/978-0-7503-5107-2ch3

3.2 The Quality Manual

Your organization may have a document called a 'Quality Manual'. The Quality Manual is a top-level summary of the quality system; it describes how your organization conforms to a standard or to a set of regulations. It used to be compulsory to have a Quality Manual, but the requirement has been dropped from some standards in recent years; many organizations still continue to write a Quality Manual as it is useful for assessors and as a training aid for staff. The Quality Manual is written and maintained by the quality management team and senior management. Its true value is in the discussions to write the document; state-of-the-art facilities and instrumentation, and investment in staff are expensive - the process of writing a Quality Manual ensures that senior management are aware and truly committed. The document may seem remote but it can bring real benefits to your career.

A Quality Manual is not a rewrite of the standard—it shows how the standard has been adopted within the organization[1]. It can be fairly lengthy, describing procedures in detail, or it can refer out to other procedures. The topics covered are summarized in figure 3.1, and further information is given in the next sections.

Foundation	• References • Definitions • Confidentiality • Impartiality • Services and products covered
People	• Structure of the organization • Staff roles and responsibilities • Responsibilities for quality system • Staff recruitment and training
Processes	• Document control • Assessing risks and finding opportunities to improve • Auditing • Purchasing • Complaints and non-conformities • Customer feedback • Management reviews
Scientific work	• Selection of methods • Equipment • Validation • Access to laboratories • Reporting results

Figure 3.1. Typical contents of a Quality Manual.

3.2.1 Foundation

3.2.1.1 References

The Quality Manual may include a list of the standards, regulations, and guidance that the organization follows.

[1] Some public organizations publish their Quality Manuals, such as the US National Institute of Standards and Technology (https://www.nist.gov/system/files/documents/2019/04/09/nist_qm-i-v11_controlled_and_signed.pdf, accessed January 2023).

3.2.1.2 Definitions
The organization will have its own jargon and acronyms—the Quality Manual is where you are likely to find an explanation.

3.2.1.3 Confidentiality and impartiality
Ethics in science is a topic that is rarely discussed but underpins everything that you do (as will be discussed in chapter 4). The Quality Manual may set out a code of conduct for staff or refer to a separate document, or ethical issues may be included in your contract of employment.

In the context of quality management, confidentiality means keeping information about one customer confidential from another—this applies to everyone in the organization including temporary members of staff. If there are exceptions to this (you may be required to report information to a regulator or to publish some results), this should be described. Who can access which documents may be described.

Impartiality is a cornerstone of quality assurance. Decisions must be based on objective evidence, the aim being to ensure that results published by the organization are unbiased. This is harder than it sounds—there can be pressure from a large customer to rush through results, and there is often temptation to take shortcuts and omit the checks that you would normally do when you are under time pressure. It can also be tempting to prioritize customers you know well, people with whom you have a personal relationship—even to include the results you know that they want. The Quality Manual sets out your organization's requirements and explains how impartiality is ensured, from the recruitment process through to on-the-job training and regular monitoring.

3.3 People

3.3.1 Organization structure

In this part of the Quality Manual, you will find a summary of how the organization is structured, its legal status, and its address.

3.3.2 Staff roles and responsibilities

The Quality Manual may include details of the different departments and the roles and responsibilities of staff within those departments.

Every employee is responsible for a quality system, but someone in your organization will be responsible for leading it. It is no longer a requirement in standards such as ISO 17025 for an organization to have a person with the job title of 'Quality Manager', but the tasks remain. The role is sometimes split between several people, and there can be a quality assurance department in large organizations, but this section will use the 'Quality Manager' title for simplicity.

The Quality Manager reports directly to senior management, feeding back any issues, particularly during management review meetings. Related to the link to senior management is the role of communicating the benefits of the quality system to all levels, helping with staff training and induction, and promoting a quality culture.

Maintaining a quality system costs time and money, and gentle reminders about why it is important are necessary.

The Quality Manager sees reports on all complaints and nonconformities, ensuring that actions to address the problems are followed through. The job also involves proactively seeking feedback from customers and other stakeholders on quality issues; this helps generate ideas to improve the quality system.

Policing the quality system is a major part of the role: the job involves arranging internal audits (audits that are carried out by colleagues) and coordinating external audits (audits carried out by accreditation bodies or regulators). This work includes following up on the findings from audits to check that any corrective actions have been completed.

In some organizations, the Quality Manager has to approve technical procedures and reports. The Quality Manager's signature is to confirm to auditors or regulators that the document complies with the quality system—the people writing and reviewing the document are approved to do so and have the right skills, qualifications, and experience; the document is formatted correctly, contains all the information needed, and was written following the quality system. The Quality Manager rarely comments on the technical content; the aim is to show to auditors and regulators that there has been some oversight.

3.3.3 Staff recruitment and training

A quality system impacts how you work as well as what you do—there is more oversight, more checking, and more record keeping than you may be used to.

It starts during the recruitment process—the organization will have a procedure to follow so that recruitment is not carried out ad hoc with a different approach for different cases. There will be a detailed job description setting out the role and responsibilities, and a person specification which includes the skills, qualifications, training, and experience needed for the role.

When you join an organization, you will follow an induction process, an introduction to the work and the workplace. No matter what qualifications and experience you have, you may have to be authorized by your supervisor or another member of staff to carry out the work (the department will keep a record of who is authorized to do what). You may first be authorized to carry out a procedure under supervision, then to carry it out unsupervised; next you may be allowed to modify the procedure and finally to develop new procedures. Moving from one level to another can take time and may involve participating in laboratory proficiency test exercises, but it is all necessary to demonstrate that you understand how the technical work is implemented in your organization and to reduce the risk to the organization and to you.

The quality system also requires that you show that you are maintaining your skills and knowledge. This is usually partly by continuing professional development (it is normal to have an agreed training plan) and partly by occasional oversight of your day-to-day work (your supervisor or a colleague may witness you carrying out a procedure to check that you are following it exactly). My experience is that witnessing of measurements comes as a culture shock—it is very different from working in academia—but it is the only way that an organization can prove to an auditor that staff are competent and are really following the procedures that the auditor has reviewed. It is all too easy to take shortcuts due to familiarity, and the steps may be there for a good reason. For example, when mixing solutions in a chemical analysis, it is tempting to cut down the time spent, but the solution can be inhomogeneous as a result.

Continued participation in laboratory proficiency test exercises is another method that is used to demonstrate your competence. Maintenance of skills and knowledge is also usually discussed during an annual appraisal of your work with your supervisor.

As with all aspects of a quality system, careful records will be needed so that there is evidence to present to auditors. This usually includes an up-to-date job description (as the job will evolve), your CV, your training record, and your training plan.

This may seem burdensome, but there are significant compensations. You are not alone: when something goes wrong you do not have to shoulder the blame all by yourself. The organization has some responsibility for not having adequate support systems in place; your supervisors must also share some responsibility. This should lead to a blame-free culture, where the important issue is to resolve the problem and prevent it happening again—this should mean you can speak freely about issues and help to find solutions, without fear of losing your job. The second advantage is that the organization is expected to invest in you, to ensure that you receive all the training you need for your job and that you stay up to date. This can only be good for your career. The advantage for the organization is that it has a skilled workforce which contributes to improving the organization.

3.4 Processes

3.4.1 Document control

One issue that may come as a surprise moving from university to industry, healthcare, or a research institute is the emphasis put on record keeping. Documents are much more closely controlled, and your work is likely to be checked by an independent person. There are good reasons for this—documents may be audited by quality assessors or inspected by regulators, the notes may have to be checked if there is a customer complaint, and you may need to prove the dates that work was carried out for a patent application. Not all documents are 'controlled'—the Quality Manual says which documents are covered.

3.4.1.1 Day-to-day notes

Document control starts in the laboratory itself with your day-to-day notes. The principles are: first, you must record enough information to enable the work to be replicated under the same conditions; second, it must be possible to identify who carried out the work and who checked it; third, any amendments must be clear, including the date of the amendment and the name of the person responsible.

In practice, this means that many laboratories continue to use hard-bound notebooks for records. These tend to have numbered pages; to demonstrate that the notes are complete, you will be asked to use indelible ink to make your notes, and any corrections have to be noted by striking through with a single line and dating and initialling the change. Each page may be signed or initialled so that the person who carried out the work can be identified. This can be laborious, so you may find that notebooks are replaced by forms to ensure all the information needed is captured. Some organizations are adopting note-taking software (such as Microsoft OneNote) which automatically records who made the data entry and when, so it is easy to check back through notes.

A further requirement of a quality system is that calculations and data transcriptions must be checked systematically. A second suitably qualified and experienced person should check that any calculations are correct, and, if data are transferred manually onto a spreadsheet or other software for more complex calculations, that person should also confirm that there have been no errors. Whether every single data transfer or calculation should be checked, or whether spot checks suffice, is often debated with quality assessors; it depends on the risk—there may be other ways further down the process that errors would be caught; on the other hand, a pharmaceutical company cannot afford any errors, and checking every calculation, every transcription, is the norm.

3.4.1.2 Technical procedures

Another difference with academia is that you are likely to be trained to follow written technical procedures—these will have more detail than any instructions you see in university laboratories. You are also likely to be asked to write your own procedures.

In the early days of quality assurance systems in the 1980s, the content needed for technical procedures (also called 'standard operating procedures') was unclear. Many early procedures I saw (and wrote) ended up as a complete description of the instrumentation and a review of the technical literature on the technique—the procedures stretched to many pages including the underpinning mathematics.

This was the wrong approach for a technical procedure; there is no need to explain any of the background information. It may be useful to record the technical background in the initial validation studies to justify the approach, but it is very difficult to maintain in the long term and difficult for someone else to follow. For a technical procedure, 'less is more'—it should contain simple step-by-step instructions you can pick up and follow easily, written in the imperative mood ('press this button…', 'adjust this setting until…', 'add this reagent to the solution…'). The usual style is numbered paragraphs, so that it is possible to identify any changes easily and to refer to other sections if needed (although I would recommend avoiding referencing other sections if possible; it is very easy to forget to change references). If the procedure includes following a particular ISO or other published standard, you may be able to make the procedure very simple, just the issues relevant to your implementation, and refer out to the standard.

There is no strict requirement for the procedure to be written instructions—you can have a series of photographs showing the instrument settings, use simple diagrams or flowcharts, or even embed a short video with a voiceover.

However, there are some points that must be covered for the quality system. The procedure normally should be checked by an independent person and approved by someone else (usually the head of department). It must have a unique document number, issue number, and date so that colleagues know they are using the latest version, and it must include a history of any changes. You would usually set out the purpose of the procedure, its scope (what it covers and what it doesn't cover), the principles, and who is responsible for what. It must be possible to show that the procedure is complete (pagination in the form 'page x of y').

The procedure must contain enough information that it is easy for a suitably qualified and experienced person to follow; inadequately documented methods can be so difficult to follow that it is impossible for someone else to replicate the work. The procedure should explain any jargon and definitions, the equipment, facilities, and laboratory consumables needed, any reference materials or data required, and any intermediate checks to be carried out.

One temptation in writing a procedure is to write down all the steps to follow if something goes wrong. This can get very long and complicated for a complex procedure—you will never finish. A workaround is to include statements such as 'if xxxx happens, refer to the laboratory manager, follow their instructions, and note the actions taken in the laboratory notebook'.

Finally, it is important to keep the procedures up to date. A procedure that includes a photograph of equipment that has long since been consigned to the bin is difficult to explain to an auditor. Methods and technical standards are constantly being improved, and procedures may need to reflect the changes.

3.4.1.3 General document control issues

As well as the work involved in maintaining day-to-day technical records and updating procedures, organizations have procedures for controlling all documents associated with the quality system. The approaches can vary from sophisticated document control software packages which automatically route documents for electronic signature to a set of folders for scanned, signed, documents with a strict file naming convention. The Quality Manual may also specify for how long documents must be retained.

3.4.2 Procurement of goods and services

Operating a quality system also means that purchasing must be controlled. This is a good thing from a scientific point of view, as it means that the technical aspects of any purchases must be considered—the purchasing department cannot simply buy the cheapest irrespective of whether it will work or not (this does happen). A purchasing procedure also helps ensure that there are checks to prevent favouritism towards a particular supplier.

The scope of a purchasing procedure covers anything that could impact the scientific work, including equipment, consumables, instrument maintenance, calibration services, reference materials, and laboratory infrastructure. The policy for procuring external scientific services may also be defined—for example, only using services that have been accredited to a specific standard.

A purchasing procedure is closely related to validation. It will include a requirement for a detailed, clear, technical specification, written by a suitable qualified and experienced person (the 'Design Qualification' stage). The purchasing procedure and the validation process work together to provide potential suppliers with a complete description, including issues such as the delivery time wanted, warranty, after-sales service, and maintenance.

If your organization has a purchasing department, it may take over at this point to collate offers from suppliers, with your help in identifying possible companies. The purchasing department looks after issues such as terms and conditions of purchase and payment schedule. Evaluation of the offers is then a joint effort, your task being to check whether the offer meets the technical requirements.

The complexity of the procedure will depend on the risk to the organization, particularly the cost. There is normally a very light touch for cheap items purchased often.

Once the item is delivered, your role is then to confirm that it is fit for purpose. For major equipment purchases, this can be related to the 'Installation Qualification' step in the validation process. As full validation can be time consuming, it may not be possible to complete all the checks quickly, and it may be necessary to confirm that payment can go ahead based on a more limited set of tests. It is a question of the risks involved.

3.4.3 Auditing and assessments

Technical work is likely to be checked by external assessors from an accreditation body or by your colleagues from another department (the terms 'assessment' and 'audit' are used interchangeably to refer to these checks). Your work could also be subject to inspections by regulatory bodies. The Quality Manual will set out your organization's policy.

The aim of an assessment is to help you and your organization. As I know from having participated in many assessments on both sides of the table, a lot of stress can be avoided by looking at the big picture and seeing an assessment as an opportunity to get feedback from others in the field. You may not agree with all of the feedback, but 'the assessor is always right', as either there was a real issue or you didn't communicate clearly enough. Both are valuable feedback to learn from for the next time—the assessor's role is that of a critical friend, and it is better to address issues than to deal with customer complaints. You may have the opportunity to take on the role of an assessor yourself during your career.

For an external accreditation body, the assessment team will usually consist of a Lead Assessor, who is an expert in quality systems, supported by one or more Technical Assessors. All of the assessors have to sign confidentiality agreements and declare any conflicts of interest, so you can talk freely to them without worrying that any information will reach your competitors. Assessors are allowed access to all relevant staff, facilities, and records, but everything is confidential unless there is a legal requirement.

Although the Technical Assessors will be experts in the field, they are not allowed to advise you on what to do, as this would be consultancy rather than assessment; equally, the Technical Assessor is not allowed to insist on something being done a particular way because they prefer it. It is up to you to show the assessor how you comply with the quality standard or regulation.

Assessments follow the same general pattern—there is an ISO standard for the assessment itself and accreditation bodies publish their own guidance. First, you may be asked to provide some documents to the assessors in advance. On the day of the assessment, you may be asked to attend an opening meeting with senior management and the assessment team. The Lead Assessor will chair the meeting, explain the process that will be followed and the assessment criteria (which standards or regulations will be used), and confirm the plan for the visit. The outcome of previous audits, changes to the organization, type or volume of work, and customer feedback may also be reviewed. The assessors will also confirm that everything discussed is confidential and that they have no conflicts of interest.

The assessment team will usually then split up, leaving the discussion of management aspects of the standard to the Lead Assessor and senior management. The Technical Assessor will join you in the laboratory to review technical aspects of the standard. Every assessment will be different, but typically the Technical Assessor will go through:

- The procedures you are following
- Results from validation studies
- Uncertainty budgets
- Results from participation in proficiency test exercises
- Equipment records including calibration certificates and control charts
- Handling of test items and artefacts
- How you demonstrate traceability
- How you report results to customers

The Technical Assessor may select one piece of work you have completed and ask to see all the records associated with this task, or they may choose one aspect of the quality system (the calculation of uncertainties, for example) and ask to see all the relevant records.

The Technical Assessor is also likely to ask to see you carry out a particular procedure, to check that the written procedure is what you do in practice and that you understand it. This should be done with some pragmatism—if a stage in the procedure involves waiting for measurements to stabilize or a chemical separation to finish, you just explain what happens and move onto the next stage. This is called a 'witnessed assessment'; as well as being used by assessors, it is often used by supervisors to provide evidence that you can be authorized to carry out the procedure unsupervised.

The general laboratory environment—the space you have to work, the layout, temperature and humidity control, in fact, anything that could affect the measurements—will also be considered.

During your discussions, the Technical Assessor should raise any concerns with you they have about any aspects of the observed work that don't conform to the standard or the regulation. This gives you an opportunity perhaps to clarify what you said or to provide more evidence, or it may be that something does need improvement.

Once all the discussions are finished, the assessment team will have a private meeting to go through the outcomes and prepare a draft of the 'Improvement Action Report'. This stage of the assessment concludes with a closing meeting which you may be asked to attend.

The closing meeting includes a brief summary from the assessment team of what they have covered and usually includes examples of good practice that they have seen (the assessment does not just identify negative aspects!). There should be no surprises for you when the assessment team gives its report; any findings in your area of work should already have been discussed with you.

The assessment team will go through all the findings from the audit—each finding has to be linked to a particular clause in the standard or the regulation, and objective evidence should be included. Each finding is classified; different terminologies may be used, but classifications are along the following lines:

- A major nonconformity—something that does not conform to the clause in the standard and could or does affect the results for customers
- A minor nonconformity—something that does not conform to the clause in the standard but does not directly affect the results for customers
- An observation—a minor deviation from a procedure or a minor oversight (forgetting to sign a report, for example)
- A recommendation—an opportunity for improvement

Nonconformities may also be classified as 'mandatory'—action must be taken to address the issue—or 'recommended'.

Each finding is then discussed (perhaps there has been a misunderstanding that can be resolved?) and actions agreed to address the findings. For a major nonconformity, the assessors are likely to ask for an analysis of the root cause of the problem, what can be done to eliminate it, and a report on the changes introduced. A time scale will be agreed for the work to be done—the assessors can ask for written evidence that the work has been completed. You may be involved in the work to clear the assessment finding.

After the closing meeting, the assessors will complete the report including the actions agreed. Their report will include a recommendation that depends on the purpose of the assessment—it could be a routine assessment to confirm accreditation or an initial application for accreditation. The assessors don't make the decision themselves; this is made by staff at the accreditation body who ensure fairness in the system by comparison with other organizations.

Techniques for assessment have been transformed in recent years. Web-conferencing software allows meetings and interviews to be held remotely, sharing documents is easy, and it is even possible to witness procedures in real time using a webcam. These techniques are often quicker and easier for the assessors, but there are risks as it is possible to miss issues. Accreditation bodies publish up-to-date guidance on the use of remote techniques and often require a balance of on-site and remote assessment techniques depending on an assessment of the risk (an organization that has held accreditation to a standard for many years may require less on-site assessment). The balance also depends on the type of assessment—informal assessments when an organization is first applying for accreditation can be carried out largely remotely.

3.4.4 Complaints and nonconforming work

Mistakes happen. Equipment develops faults, air-conditioning systems drift outside specification, people make errors when following procedures, or the written procedure may not match what is being done in practice. It is inevitable that work will be carried out that does not follow the quality system exactly—this is called 'nonconforming work'. A nonconformity can be spotted by a member of staff, perhaps when the results are being checked by a colleague, or identified during an audit. You may also receive a complaint from a customer who has found a problem with the results.

It's important that any nonconformities or complaints are investigated and the causes identified. Actions must be taken to deal with the immediate issue, check whether there have been other examples, and prevent a recurrence. The Quality Manual will describe a formal process to follow, which usually includes a to capture all the information needed. The advantage of following a formal process is that it should enable problems to be reported with a focus on the issue rather than seeking to blame individuals. Reporting a nonconformity should not be seen as a fault. There is a parallel with health and safety issues—some organizations encourage near misses to be reported so that action can be taken before there is a serious accident.

The report is then reviewed, its significance assessed, and corrective actions decided. The immediate actions could involve repeating the work and withdrawing results, or even stopping the service while the problems are resolved, while ensuring that anyone affected is informed (including the accreditation body or the regulator, if necessary). Then come actions to address the root cause—repairing or replacing equipment, changing the procedure, improving training, etc—and authorization to restart the service if it has been stopped.

Nonconformity reports are collated and used to identify general opportunities to improve the quality system, so the organization benefits in the end.

3.4.5 Customer feedback

Many quality systems include proactively seeking feedback from customers by sending out a form for them to complete or by sending a link to an online form. It is normal for the completed form to be returned to an independent person, as the customer may know you personally.

Typically, customers are asked to rate their satisfaction with various aspects of the service, including whether the service met their expectations for the results, measurement uncertainties, clarity, turnaround times, etc.

The return rate on these forms can be low—if 10% of the forms are returned, it is quite good. Despite this, seeking customer feedback is beneficial. First, even if the customer does not respond, it shows that your organization is focussed on their needs. Requesting feedback may encourage some complaints, but it is better to know that the customer is unhappy so you can try to address their concerns rather than losing a customer or hearing about the complaint through someone else. Much of the feedback will be positive, and the forms are an opportunity to thank colleagues for their hard work.

3.4.6 Management of risks and opportunities

Risk reviews are common in the context of health and safety, but the requirement for a process to address risks and opportunities is relatively new for quality systems (it first appeared in ISO 17025 in the 2017 edition). It recognises the reality that an organization cannot control everything in minute detail—there are always risks—but requires an organization to consider what the risks could be. It is a very useful addition as it allows an organization to recognize that a risk exists and to take action, or to accept the risk if cost and other factors are important, all in a transparent way. This can lead to better decision making and helps quality assessors and regulators have confidence in the organization.

In practice, organizations have to plan and implement actions to address risks or improve the quality system. If you have already sat through health and safety risk reviews, you will know that they can be fairly dry meetings as you are discussing what could happen—the emphasis in the quality system is on the actions that emerge, not on the process itself.

Risk is a combination of the probability of the occurrence of harm and the severity of that harm. Many different techniques are available for assessing risk; sophisticated software packages exist that evaluate the probability and severity based on a number of factors, reevaluate the risk for different actions, and output priorities.

3.4.6.1 A simple approach to a risk review

The aim of a risk review is to identify what could go wrong, determine the impact if it does go wrong, and decide the actions to take and their priority. A simple approach to a risk review can be effective as it is the actions that are important, not how they were decided.

The first step is to work systematically through the procedures from beginning to end, reflecting on what could go wrong. Typical problems could be equipment breakdowns and faults, cross-contamination of samples, issues with utilities (for example, air-conditioning faults), training of new staff, data transcription errors, and potential errors in setting up equipment. One can also consider customer feedback, assessment reports, nonconformities, any changes to the work, changes to staff, organizational changes, technical data such as control charts, results from Proficiency Test Exercises, reviews of scientific literature, external reports, and feedback from conferences. Risks to other aspects of the quality system, such as confidentiality or impartiality, should also be covered. A fishtail diagram can be useful to identify possible problems. Quality-critical industries such as pharmaceutical producers and aerospace use detailed fault-analysis approaches to identify possible problems, such as fault-tree analysis.

The next step is to consider the consequences of the possible problems. What would be the impact on results, measurement uncertainties, or traceability? How many customers would be affected and what impact would it have on them; would the organization lose customers or reputation as a result? Are there any regulatory issues, or could accreditation to a quality standard be affected?

The simplest way to prioritise the risks is to decide whether the probability is high or low and whether the impact is high or low. Other risk analysis systems have more classifications —'very low, low, medium, high, very high', for example. Whichever system is used, high priority is given to address risks that are high probability and high impact, low priority is given to low-probability and low-impact risks, with the other categories in between.

The actions that can be taken include:

- Do nothing—accept that the risk exists but is at a level that the organization can accept, so the risk would be kept under review and action taken if something changes
- Take actions to reduce the probability—change the procedure, purchase duplicate equipment, automate work to avoid data transcription errors, introduce new intermediate checks, set up new training…
- Take actions to reduce the impact—account for variations in the laboratory environment in the uncertainty budget, set a longer lead time for publishing results to allow time for repairs or repeating work…
- Develop a contingency plan, so that actions are already decided if the risk comes to pass—arrange with another group to borrow equipment if something breaks down…
- Outsource the risk—take out insurance
- Pass the risk up the chain for senior management to consider

As part of this review, you will probably identify opportunities to improve ways of working—actions you can take to improve compliance with the quality system but that aren't concerned with hazards that can harm results. These opportunities should also be noted.

The final, and the most important, step is to carry out the actions you have decided and communicate your decisions to colleagues and other stakeholders. This means documenting the actions and embedding them in forward job plans or project plans, so that it is clear who will carry out the actions and by what date. Otherwise, conducting a risk review is a meaningless process.

3.4.7 Management reviews

The Quality Manual may also describe how senior management checks the health of the quality system. An annual meeting is common—the Quality Manager may ask you for a contribution to a report that is being prepared for the meeting, or ask you to attend the meeting yourself.

The management review covers all the issues that impact the quality system, including:
- Changes in customer requirements
- Changes in resources (staff, equipment, laboratory infrastructure)
- Changes in the volume and type of work, new services launched, and services withdrawn or suspended
- Customer complaints and feedback
- Feedback from internal and external audits
- Plans for future audits
- Trends in nonconformities
- Progress on actions from the previous management review
- Status of actions from nonconformities, complaints, and audits
- Suitability of the Quality Manual and management procedures
- Feedback from risk reviews
- Opportunities to improve the quality system
- Staff training needs
- Actions to promote the quality culture

The management review is an important event in the quality system calendar, as it can bring requests for investment in new equipment, infrastructure, and staff to the attention of senior management.

3.5 Scientific work

The Quality Manual will also give an overview of how the scientific aspects of the work comply with the standard or regulation. There will be a top-level summary of the principles set out in chapter 2 of this book, so it is a useful place to get an idea of how the laboratory work is structured.

References

[1] 2017 *General Requirements for the Competence of Testing and Calibration Laboratories* ISO/IEC 17025:2017 International Organization for Standardization

[2] 2015 *Quality Management Systems—Requirements* ISO 9001:2015 International Organization for Standardization

IOP Publishing

Quality Assurance for Scientists and Engineers
A practical guide
Steven Michael Judge

Chapter 4

The future of quality assurance

4.1 Digitalization

One might be forgiven for thinking that there is nothing more to do in the field of quality assurance, other than the challenges of implementation and gradual improvement. The principles and procedures are well established, standards have been refined over many years, and work that was pioneering a few decades ago is now just routine in many organizations.

Nothing could be further from the truth. We are at the beginning of the fourth industrial revolution, driven by advancements in information technology; the factory of the future will be fully integrated and automated, with more robotics and more dependence on software, artificial intelligence, and machine learning. The A*STAR demonstrator factory in Singapore[1] shows what the future could hold, with all the equipment and instrumentation monitored in a central control room and staff training carried out using 3D digital twins of the equipment.

In response to these changes, all the building blocks of the quality infrastructure are likely to be transformed in the coming years. A report in the 2021 bulletin from the International Organization of Legal Metrology [1] sets out a vision of a new digital quality infrastructure, including innovative instrumentation and integrated databases. The changes are happening both bottom-up and top-down, with individual organizations adopting new approaches and international networks, such as the International Committee for Weights and Measures (CIPM) and EURAMET (the European regional metrology organization), seeking harmonization.

Digitalization is defined to be 'the use of digital technologies to change a business model and provide new revenue and value-producing opportunities; it is the process of moving to a digital business' [2]. This means different things to different people, but the first ideas are already visible.

[1] The demonstrator factory can be seen at https://www.a-star.edu.sg/simtech/model-factory-at-simtech/overview.

doi:10.1088/978-0-7503-5107-2ch4

4.1.1 Machine-readable reports and calibration certificates

The starting point for digitalization of services is likely to be changes to calibration certificates, which will impact organizations that issue certificates and organizations that use calibrated instruments (i.e., all scientific establishments). Traditionally, calibration certificates were printed on watermarked paper, signed by the staff responsible, and posted to the customer to emphasize the value of the certificate. The first step on the digitalization journey was replacing the paper certificate with a scanned version, emailed directly to the customer. The next step is a document with a digital signature, providing a higher level of security as the certificate is encoded. In parallel with this, organizations moved to transferring the information automatically to certificates rather than having a separate step in which data were copied across into a Word document for printing. In these examples, the certificate reaches the customer faster, but the customer still has to transfer the information to the software package used in the laboratory.

The obvious next step is to make the certificates machine readable to eliminate the risk of human error, so that the information can be extracted into the instrument software without manual data transcription. The International Bureau of Weights and Measures (BIPM) has implemented an example of this technology, and machine-readable comparison reports are being launched (see case study 4.1). Agreement on the format and content of certificates is essential for this to be implemented. Work on this aspect is at an early stage, with groups such as EURAMET developing clear guidelines. A unified system of metadata (an 'ontology') is needed so that all digital certificates include the numerical value, SI unit, measurement uncertainty, coverage factor, and probability density function in the same format—the 'Digital SI' is under development by a working group of the CIPM and more will be heard of this in the years to come[2].

Case study 4.1 Machine-readable reports [3].

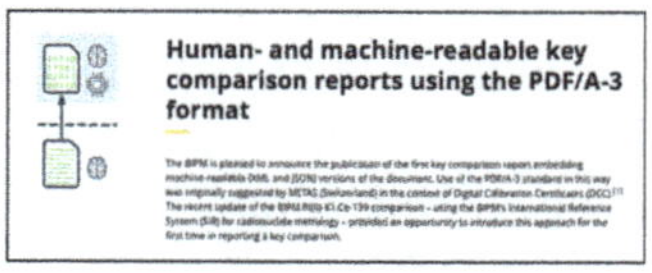

Comparison reports issued by the BIPM are being made available in PDF/A-3 format. Reports can be read by a computer, and it is possible to include rich metadata such as administrative and technical details. This makes it easier for metrology institutes to extract their results without manual data transcription; it also enables more detailed studies of the results of comparison exercises than were previously practical, as the effect of any parameter on the results can be studied.

Dr Romain Coulon, BIPM, Paris 2022

[2] Reports on progress will be available on the BIPM website: www.bipm.org.

A true digital calibration certificate can hold more information than the paper version. Some calibration certificates can be very complex; a digital certificate can include lengthy tables and plots that are difficult to fit in a paper version. Additional information can be embedded, linking the data back to primary standards. Some machine-readable information on calibration is already available; it is possible to extract machine-readable information on the calibration and measurement capabilities of national metrology institutes from the 'Key Comparison Database' on the BIPM website using an application programming interface (API).

4.1.2 Confidentiality and security

Digitalization will make data more widely available, but there are implications for the confidentiality and security of the information. There is a risk that sensitive information could be accessed and used by competitors.

Case study 4.2 emphasises how new tools such as blockchain can be used to improve data security.

Case study 4.2. The impact of digitalisation: the view from an accreditation body. A major driver for accreditation and mutual recognition of quality assurance systems is enabling trade, ensuring that a product manufactured in one country is accepted as meeting the standards of another country.

Digitalization has the potential to make inspection of shipping documents at borders much easier and faster. Evidence of conformance to specification, maintenance records, and much else besides can be captured in a digital record. If the information is in a secure and harmonised machine-readable format, the documentation may even be checked by an artificial intelligence system.

The United Kingdom Accreditation Service (UKAS) is making the first steps towards a digitalised quality infrastructure. UKAS-accredited laboratories are issued with a unique QR code that is linked to a secure database of accreditation certificates. The system uses blockchains so that a permanent record of any changes to the data is retained, preventing unauthorised changes. The aim is to prevent fraudulent claims that a laboratory has been accredited. The system (called E-certificates) is being expanded to cover details of the services covered, so that customers can confirm that a laboratory has been accredited for a particular service.

There are challenges to widespread adoption of the new digital technologies. Some of the information may be confidential and have commercial implications, so international agreement is needed on what a digital record would include. Some conformity assessment bodies are relatively small and have limited resources to implement digital platforms, so introduction will take time.

There is no doubt that digitalization will transform quality assurance in the years to come, with many opportunities for scientists to contribute. This does not mean that everyone in the field must become a software engineer, but it will be important to understand the principles of the new technologies and to stay inquisitive about the possibilities.

Jeff Ruddle, Strategic Development Director, United Kingdom Accreditation Service
Interviewed by the author in December 2022

4.1.3 The instruments of the future

We are familiar with instruments in which the information obtained from a sensor is presented to us on a built-in display or captured and processed by a computer. The instrument of the future may be just the sensor: all the data and processing software would be held in the cloud—everything would be virtual except for the sensor itself. This would transform legal metrology, as it would enable regulators to review all the documentation for an instrument, see the results, check performance data, and look for nonconformities, simplifying some of the cumbersome procedures that must be followed in this domain.

Other developments in instrumentation include harmonizing data formats so that it is easy to share information. One example is the design of instruments to detect radiation dose rates and radioactivity on surfaces; in the event of an incident such as an accident at a nuclear site or the detonation of a 'dirty' bomb in a terrorist attack, the emergency services are called upon to go into areas that could have high dose rates or radioactivity. They are issued with basic radiation detection instruments, but interpreting the measurements can be difficult. A recent International Electrotechnical Commission (IEC) standard [4] describes a common data format for such instruments to use so that the measurements can be transmitted back to experts in a central facility to guide the work. The IEC standard can be used by manufacturers when designing new instruments.

Instruments of the future could also break the model of traceability and equivalence completely. The assumption of the current international measurement system is that primary measurement standards are complex and need specialist expertise to realise, so the most efficient way to maintain a harmonised measurement system is to have the primary instruments and expertise concentrated in a single laboratory. The US National Institute of Standards and Technology's 'NIST on a Chip' programme aims to revolutionise the field, enabling primary standards to be realised in any laboratory from atomic clocks through to magnetic and electric field sensors (see case study 4.3). There would be no need for such instruments to be calibrated; the link to the definitions of the units would be inherent in the design.

Case study 4.3. The instruments of the future: NIST on a Chip.

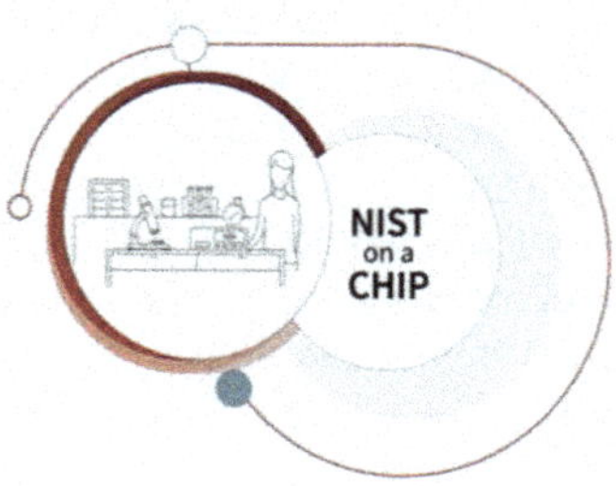

The US National Institute of Standards and Technology's 'NIST on a Chip' programme heralds a paradigm shift in scientific instrumentation[1]. The vision of the programme is that instruments will be their own primary standard, and the concept of an unbroken chain of calibrations back to a national or international standard would be superseded. The new instruments will be rugged, cheap, and more accurate, thanks to developments in manufacturing techniques for integrated circuits.

The origin of the programme was the pioneering work of NIST physicist John Kitching, who worked with a team of scientists to develop a compact atomic clock orders of magnitude smaller than the primary standards used at national metrology institutes. The project ultimately resulted in a highly accurate, low-power atomic clock, the size of a typical integrated circuit. Working in partnership with private industry and with support from the defence sector, more than 100 000 of these miniature atomic clocks have been sold. Applications include improved GPS systems, underwater sensors for gas and oil exploration, laser ranging, and the control of unmanned aerial vehicles.

The team's work showed that 'NIST on a Chip' was feasible. The 2019 redefinition of the SI base units in terms of fundamental constants of physics opened up new possibilities for other primary sensors based on the principles of quantum mechanics. The ambitious new programme envisages the development of ultrareliable, compact quantum sensors for time, frequency, distance, mass, force, temperature, pressure, electrical current, voltage, magnetic field strength, etc—and these sensors can be combined to measure derived quantities or be used in networks to carry out measurements over large areas. The new sensors will represent the first applications of quantum technologies in day-to-day life, impacting everything from medical diagnostics to transport and manufacturing (Industry 4.0), and all without the need for calibration.

This ambitious programme is not without challenges. In some fields, the underpinning technology does not exist yet: for example, the miniature lasers needed for electric field sensors are still under development. The question of quality assurance of the new devices is also difficult; primary standards could be produced and disseminated by commercial companies rather than being the sole providence of national metrology institutes, and periodic recalibration will not be needed. Without evidence of traceability to an independent national or international standard, what changes to quality assurance standards will be needed so that auditors or regulators will be confident that the instrument is operating correctly? However, the potential benefits of the new quantum devices are so clear that these problems will be solved, and other national metrology institutes are also involved in the field as a route to transfer their knowledge and expertise to industry [5]. The International Measurement Confederation, a group of 42 organizations that encourages cooperation among scientists and engineers working on metrology and instrumentation, has recently set up a technical committee to share information on developments[2].

The NIST on a Chip programme shows how quantum technologies will transform instrumentation and our world. For scientists, this will offer skilled jobs in the 'quantum workforce' that will be needed to bring the ideas to reality.

Barbara Goldstein and Jay Hendricks
National Institute of Standards and Technology, USA

[1]More information on the programme can be found on the NIST website: https://www.nist.gov/noac.
[2]More information on the International Measurement Confederation is available at https://www.imeko.org/index.php/about-imeko.

4.1.4 Documentary standards of the future

Documentary standards, such as those published by the International Organization for Standardization (ISO) and the IEC, are written by a team of experts, usually during face-to-face meetings, and then follow an approval procedure before publication. The process can take several years, and any changes to the document can also take one or two years. Once the standard has been published, it is then up to the user to interpret and implement the requirements; any equations and data embedded in the text have to be entered into spreadsheets or software for use in the workplace. This process of transferring knowledge from the expert group to implementation is robust and tried and tested, but several years can elapse between the original ideas and their use.

The Digital Standards Initiative (IDiS) [6] aims to transform this picture [7]. The vision is that standards will be machine readable[3], implementation will be automatic, and artificial intelligence will be used to update the standards themselves. IDiS has identified five different levels that are theoretically possible within ten years (figure 4.1).

Level 1 has already been achieved: most standards can be downloaded in digital (PDF) format. There are already concrete moves towards Level 2 digital documents. The main improvement is that it will be easier to create and distribute content; standards bodies are adopting online collaborative document systems so that all the authors can work on a standard at the same time, rather than the text being discussed

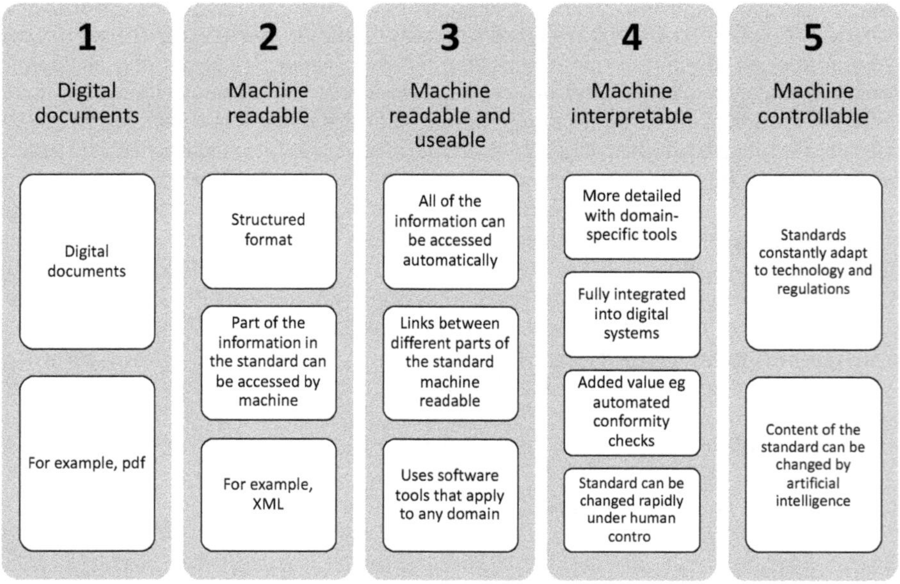

Figure 4.1. Scenarios for the future of documentary standards.

[3] See, for example, https://www.iec.ch/blog/making-standards-machine-readable.

during meetings[4]. Some parts of Level 2 documents would store metadata, so some elements of the text and the structure of the document would be machine readable.

Level 3 documents would be entirely machine readable. Instead of being written in prose that has to be interpreted and implemented by the reader, standards would be broken down into small chunks of information with associated metadata that could be interpreted by a machine and used for process control and automation. Some digital objects—equations, reference data, etc—could be used directly or exported to users' systems (for example, equations could be transferred into spreadsheets or mathematical software packages with no user intervention, reducing the risk of errors).

At the Level 3 stage, the tools available to use the information in the standard would be generic; Level 4 documents add domain-specific software tools. Information from users can be fed back to guide the creation of the next version of the standard.

The final level, Level 5, is a fundamental change in documentary standards. Consensus-based standards that rely solely on experts in the field would be replaced by automated artificial intelligence decisions. Standards are no longer static but constantly evolve, with artificial intelligence programmes detecting the need for change, based on knowledge collated from machines, social networks, and experts, linked to regulations. The standards become machine-controllable.

4.1.5 Ensuring the quality of artificial intelligence and machine-learning algorithms

One challenge for the future will be how to validate software that uses artificial intelligence or machine-learning algorithms, as the software does not follow a step-by-step process but learns from training data. This is far from being a purely academic issue: for example, for nuclear safety, environmental protection, and nuclear security one often needs to identify and quantify the radionuclides in a sample using a spectroscopic technique. Analysis can be complex, taking into account information from the whole spectrum, considering the complete decay scheme of a radionuclide, and rejecting artefacts that can be confused with a signal–it is a mix of logic and intuition gained from experience of analysing hundreds of spectra. It would seem to be an excellent example of a problem that could be solved using an expert system, a neural network, and several groups have attacked the problem [8]. However, the new approaches are yet to be accepted by the community (it is not included in the current ISO standard for the technique), and measurements continue to rely on validated software and trained staff, although an expert system might save significant time and cost.

This caution is probably justified; one research group has identified 31 problems in the quality assurance of machine-learning algorithms, ranging from bias in the data used to train the software to the difficulties in writing a specification [9]. It can be difficult to predict exactly what the programme will do, and hard to reproduce

[4] See, for example, the Online Standards Development (OSD) platform by ISO and IEC: https://www.iso.org/OSD. More information about OSD is available at https://sway.office.com/s4WjFoQKazarZFpF?ref=Link.

results. The long-term stability of the code is also difficult to validate. Nonetheless, given the potential benefits there is significant research underway to find quality assurance methods.

4.2 Quality assurance and academic research (the reproducibility crisis)

A famous article 'How science goes wrong' was published in *The Economist* in 2013, questioning the reliability of the results of academic research [10]. The article picked up on a long-standing debate in the academic community about the difficulty of replicating the work of other groups, casting doubt on the value of scientific research.

Many reports and articles have been published on this topic, including a comprehensive review by the US National Academies of Science (NAS) in 2019 [11][5]. A survey published in *Nature* [12] found that more than 70% of researchers across different disciplines encounter difficulties in replicating the work of others (the survey was self-selecting so the results were probably biased, but nonetheless the article showed the ongoing concern).

There are strong incentives to solve the problem. As the NAS pointed out, the results of scientific research can impact government policy, including contributing to economic growth, healthcare, and environmental protection. Research is not done in isolation; the impact of an error can be multiplied many times when other teams across the world or in the future base their studies on faulty information. Time spent puzzling out why results are different is time not spent advancing science—with funding for research in short supply, the community must aim to make the best use of the money that is available.

Several root causes have been identified ([11–15]):

4.2.1 Cause #1: working practices

As an example, many years ago, I was setting up a magnetic spectrometer for the next series of measurements when the head of department burst into the laboratory and frantically scrabbled around in old printouts of spectra. The background to this came out later—there had been an accusation that a student had falsified data by adding artificial peaks to spectra. Sure enough, the peaks were not present in the raw data and had been added by the student.

Deliberate falsification of data such as this is very rare, but this personal anecdote captures many of the issues identified by the NAS. When one thinks about the root causes, one can perhaps have some sympathy with the misguided student or at least understand why they behaved like that. First, undergraduate and graduate science courses focus entirely on the subject matter; there is little or no guidance on the reality of working as a research scientist. The work tends to be unstructured and informal, with little detailed peer review or mentoring.

[5] The NAS report uses the term 'replication' to refer to repeating the experiments of another research group and 'reproducible' to repeating one's own work.

In the past, perhaps this worked because research was conducted at a slower pace (there was no choice, data collection and analysis were by hand, communication between groups was by letter), and students had more time to learn a public service ethos in research from their supervisors. Ensuring the quality of work is rarely covered in university courses—the NAS's guide [16] is a rare exception to this lack of guidance.

Second, there is intense pressure to publish, coupled with job insecurity—the next job, the next research grant, can depend on publishing in a prestigious journal. Novel, unexpected results are those most likely to be published, yet these are probably the results that should be subjected to the closest scrutiny by the referees. On top of this, there are no prizes for publishing an important research finding second; the high-profile journals are less likely to publish confirmatory research. The temptation can be overwhelming to leap to conclusions or to cut corners by not checking the data carefully. There is always a risk of bias to meet a publishing deadline or complete a thesis before the grant runs out. None of us is immune—I suspect we have all used the phrase 'A typical plot of the results is shown in figure 1', meaning here is the most convincing plot from the set we had available.

Related to the pressure to publish is the third issue: there can be a lack of time to focus on the hard graft, the '99% perspiration' of research. The NAS report highlights a lack of detail in papers and reports of methods, instruments, materials, procedures, statistical methodology, and measurement uncertainties, making it difficult for other groups to replicate the work. Basic record keeping in laboratory notebooks is often scribbled incomprehensible notes (a colleague and I ran a short course on record keeping for new research scientists, and we were taken aback that the material came as a surprise to many).

Fourth, there is pressure not to challenge a colleague—it is just 'not done'. Scientists are reluctant to reach out to original researchers and to be seen to accuse them of incompetence. I fell into this trap myself in a minor way when I read a paper that assigned a spectrum peak to a new nuclear decay mode rather than the spectrum artefact that it was: to chase the author (who had already left their job) to publish a correction was a challenge I didn't want to face. Perhaps it would have been different if I'd thought the impact was high; the Royal Netherlands Academy of Arts and Sciences points out [17] that a high priority should be given to replicating studies if their impact on society or knowledge is high or if the work influences public policy or teaching.

Given that organizations and the community can value publishing over ensuring quality, the decision taken by the student to falsify the results can be understood, if not forgiven.

4.2.2 Cause #2: knowledge has uncertainty

You may have encountered scientists who take questions about uncertainties as a personal affront; 'there are no uncertainties in my work' is a phrase I've heard more than once. However, most science involves measurement, and there are always subtle variations if methods are looked at closely enough—choices of the equipment to use, the grade of reagent to select, even the method used to mix solutions can make a difference. This margin of doubt has to be characterized: uncertainties are core to science. Nothing in experimental science is certain; all one is often doing is improving confidence. Despite this, research papers often do not include detailed uncertainty budgets, and components are sometimes not recognized—this makes it impossible to check whether results from different groups are consistent.

4.2.3 Cause #3: lack of training

Chapter 2 of this book has highlighted the challenge of validating software. Instrument control, data acquisition, data management (transforming, processing, archiving), and analysis (modelling, simulation, data visualization) all depend on software, but this sits in the hinterland between science and software engineering. Software development is full of traps; numerical analysis can be unstable if the wrong algorithm is used; the outputs of nonlinear systems can depend on the initial assumptions. The NAS report identifies that training for scientists has not kept pace with the rapid changes in information technology.

Another issue often encountered is in statistical analysis, with difficulties interpreting results. There are many examples of conclusions being based on too few experiments, particularly in biomedical research where each replication is expensive.

4.2.4 What can be done about the reproducibility crisis?

The NAS identifies actions specific to the academic community. A heavy responsibility lies with the academic journals and the editors and referees, as they control the endpoint of much academic research and hence the careers of the authors; in many ways, journals are the customers for the work.

The advice includes encouraging journals to set tighter standards for papers before they can be accepted for publication. Much more detail should be included on experimental methods, equipment, uncertainty budgets, and the software used. The checks carried out on the papers should include checking calculations where possible, plotting data in different ways, and comparing results to those of other authors—similar to the checks that a technical assessor might carry out. Another approach that the NAS recommends is that data analysis plans should be approved before the experimental work starts, to reduce the temptation to select data to match the theory.

A successful example of this rigorous approach to publication is the process used in the nuclear industry to publish definitive information on the decay of radionuclides (half lives, particle and gamma emission probabilities, etc). The process (called the Decay Data Evaluation Project) sits on the boundary between academia and industry, taking data often originally measured for nuclear structure studies and following a strict process to check the data and evaluate the measurement uncertainties in detail. The outcome is a database that is accepted worldwide by the nuclear industry for work important for our health and environmental protection, and by medical physicists to ensure the safety and effectiveness of imaging and therapy (see case study 4.4).

Case study 4.4. On the boundary between academia and industry: the Decay Data Evaluation Project (DDEP).

The aim of the DDEP is to provide the nuclear community with definitive radioactive decay data such as half lives and emission probabilities. These data are needed for the safe and effective use of radiopharmaceuticals for imaging and therapy, and by the nuclear industry.

The raw data, often from academic studies of varying quality, are subjected to a critical review by experienced physicists. Measurement uncertainties are adjusted if necessary to meet the requirements of the *Guide to the Expression of Uncertainty in Measurement*. The data are analysed, following a written procedure, to obtain evaluated decay data. The physicists carrying out the work are trained on the procedures, and regular workshops are held to discuss results and seek continual improvement.

Historically, nuclear decay data were evaluated by different organizations, and databases were sometimes held at national level. The result of the DDEP is a single database that is accepted worldwide.

There have to be incentives for researchers to meet more stringent requirements for publication. Public recognition of the efforts made could be through 'badges' applied to papers and prizes for the most reproducible work. Funding for review articles should also be increased, so that there is independent oversight.

There is a caveat that because repeating work is expensive, resources should be focussed on issues that could have an impact on public policy or basic knowledge. The decision to carry out confirmatory research should take into account the risks, how surprising the result was, the potential for bias, the complexity of the work, the possibility of uncontrolled variables, and also whether there are alternative methods to confirm the result.

Estimation of measurement uncertainties is highlighted as a particular concern. Some authors point to the value of investigating uncertainties and inconsistencies between results; such studies can identify previously unidentified phenomena and inspire progress.

Software is also an issue of concern. Researchers trying to repeat measurements even a short time later find that code doesn't work, analysis libraries have changed, operating systems have been updated, and the code no longer even compiles. This is so common that I suspect that you have come across the problem yourself many times (I've even come across a case where the source code had been lost, only the executable code remained, and the department had to keep obsolete computers to run the programme). Careful documentation, version control, and the use of nonproprietary data formats are key to ensuring that colleagues can benefit from the work invested in developing the software, but this all takes time.

The issue of the proliferation of data and software is so important that we are witnessing the beginnings of a new international body. An organization called 'GO FAIR' has been established by France, Germany, and the Netherlands with the aim of ensuring that scientific data can be reused[6]. The guiding principles were published in *Scientific Data* in 2016 [18]:

- **F**indable: It must be possible to find electronic records, with detailed metadata (electronic labels) incorporated in the data, a unique ID, and registration in a recognized resource.
- **A**ccessible: Data must be formatted so that they are easy to retrieve using a standard protocol.
- **I**nteroperable: Data must use a formal, shared, language so that they can be used by different software and combined with data from other resources.
- **R**eusable: The data must be described in detail, with a clear provenance (who produced the data, when, how the data were processed, how to cite the work) so that they can be reused into the long term. The vocabulary (ontology) to use depends on the field of work.

GO FAIR communities are already being established, the first step being workshops to define the metadata to describe the data. Examples include laboratories working in the fields of wind energy and studies of infectious diseases—current examples are given on the GO FAIR website. The implications of GO FAIR are that raw data will be available to others working in the field, opening up the data themselves to closer scrutiny.

It will take some time before the measures recommended to address the reproducibility crisis are accepted and implemented in full. Standing back, one can see that the actions suggested are:

- More formal ways of working—detailed procedures, clear record keeping
- Equipment records and careful calibration
- Better control of data
- Better training
- Mentoring and peer review by 'critical friends'
- Better use of statistics and detailed uncertainty budgets
- Comparisons with other methods and other groups
- Independent audits
- Risk assessments
- Public recognition of the actions taken

The similarity of these recommendations to address the reproducibility crisis and the requirements of a quality assurance management system is not surprising—they are 'obvious common sense'. The difference is that the people calling for the changes

[6] The GO FAIR website is https://www.go-fair.org/.

are leading academics rather than regulatory and accreditation bodies. Writing in *Nature*, Sené, Gilmore and Janssen say 'If researchers are properly trained to use best metrological practice, following clear procedures and calibrating their measurements against standards that are directly linked to the agreed base units, we can all have confidence in their results' [14].

Scientists working in industry benefit greatly from their experience in academia. Perhaps it is time for this debt to be repaid, and the so-called reproducibility crisis is an opportunity to adapt practices from industry and from the field of metrology to underpin trustworthy research.

References

[1] Thiel F 2018 Digital transformation of legal metrology—the European Metrology Cloud *OIML Bull.* **LIX**(1) 10–21 (https://www.oiml.org/en/publications/bulletin)

[2] Digitalization. Gartner Glossary (https://www.gartner.com/en/information-technology/glossary/digitalization)

[3] Coulon R, Courte S, Judge S, Michotte C and Nonis M 2022 Digitalization of the reporting of key comparisons for radionuclide metrology *Meas. Sci. Technol.* **33** 024003

[4] 2018 *Nuclear Instrumentation—Data Format for List Mode Digital Data Acquisition Used in Radiation Detection and Measurement* IEC 63047:2018 International Electrotechnical Commission

[5] Tzalenchuk A, Spethmann N, Prior T, Hendricks J H, Pan Y, Bubanja V, Temporão G P, Yu D H, Ilić D and Goldstein B L 2022 The expanding role of national metrology institutes in the quantum era *Nat. Phys.* **18** 724–7

[6] Czarny D A, Diemer J, Schacht M, Bülow G, Noll M, Lochner D, Gayko J, Pervin N, Rauh P and Diedrich C 2021 *Whitepaper: Scenarios for Digitizing Standardization and Standards* DKE (https://www.din.de/resource/blob/801106/0251eb1280a9a97e53285d42d3bf1fea/whitepaper-idis-en-data.pdf)

[7] Donnellan A private communication 2023

[8] Kamuda M, Zhao J and Huff K 2020 A comparison of machine learning methods for automated gamma-ray spectroscopy *Nucl. Instrum. Methods Phys. Res.* A **954** 161385

[9] Alamin M A A and Uddin G 2021 *Quality assurance challenges for machine learning software applications during software development life cycle phases IEEE Int. Conf. on Autonomous Systems (Montreal, 2021)* (New York: IEEE)

[10] 2013 How science goes wrong *Economist* 21 October 2013 (https://www.economist.com/leaders/2013/10/21/how-science-goes-wrong)

[11] National Academies of Sciences, Engineering, and Medicine 2019 *Reproducibility and Replicability in Science* (Washington, DC: National Academies Press)

[12] Baker M 2016 1,500 scientists lift the lid on reproducibility *Nature* **533** 452–4

[13] Milton M J T and Possolo A 2020 Trustworthy data underpin reproducible research *Nat. Phys.* **16** 117–9

[14] Sené M, Gilmore I and Janssen J T 2017 Metrology is key to reproducing results *Nature* **547** 397–9

[15] Harris R 2017 *Rigor Mortis: How Sloppy Science Creates Worthless Cures, Crushes Hope, and Wastes Billions* (New York: Basic Books)

[16] National Academies Committee on Science, Engineering, and Public Policy 2009 *On Being a Scientist: A Guide to Responsible Conduct in Research* (Washington, DC: National Academies Press) 3rd edn

[17] KNAW 2018 *Replication Studies: Improving Reproducibility in the Empirical Sciences* Royal Netherlands Academy of Arts and Sciences (KNAW)

[18] Wilkinson M D, Dumontier M and Aalbersberg I J *et al* 2016 The FAIR guiding principles for scientific data management and stewardship *Sci. Data* **3** 160018

Conclusions

Working to a quality assurance management system is something that most scientists will experience in their careers. Quality management systems such as ISO 17025 and ISO 9001 are founded on best practice learnt by others over generations; they are not intended to stifle research or initiative but to help communication with end-users and colleagues, and to help you be proud of your work.

The International Organisation for Standardization has identified a set of 'Quality management principles': a set of fundamental beliefs, norms, rules, and values that are accepted as true[1]. The principles include focussing on the end-user and using evidence to guide decisions, but in essence all of the principles are about engaging with the quality system positively. The chapter on future developments shows that quality assurance systems bring opportunities for innovation.

However, the last word belongs to W Edwards Deming, one of the leaders of modern quality assurance, who said 'Without questions, there is no learning'. Quality assurance management systems are intended not to be followed on autopilot, but to help you ask the right questions.

[1] ISO 2015 *Quality Management Principles* International Organisation for Standardization (https://www.iso.org/publication/PUB100080.html).

doi:10.1088/978-0-7503-5107-2ch5

www.ingramcontent.com/pod-product-compliance
Ingram Content Group UK Ltd.
Pitfield, Milton Keynes, MK11 3LW, UK
UKHW051849100926
8556IPUK00005B/543

9 780750 351058